Cotton and Allied Textiles Industry Advisory Committee

Safety at finishing plant and machinery

CONTENTS

FOREWORD

The Health and Safety Commission has endorsed
the practical guidance contained in this document
which it commends to the cotton and allied textiles
industry.

PREFACE

The Cotton and Allied Textiles Industry Advisory Committee (CATIAC) was established in 1988 to advise the Health and Safety Commission on matters pertaining to health and safety in the cotton and allied textiles industry. CATIAC superseded the former Joint Standing Committee which had been wound up.

CATIAC agreed to complete the work already begun by the Joint Standing Committee on the preparation of guidance on safety at plant and machinery used in the finishing sector of the industry. A Working Group was formed to carry out the task and comprised representatives from employers, machinery manufacturers, trades unions and the Health and Safety Executive. The guidance so prepared and set out in this document has been fully endorsed by CATIAC.

This is an authoritative document that will be used by health and safety inspectors in describing reliable and fully acceptable methods of achieving health and safety in the workplace. It remains open to employers to achieve equivalent levels of health and safety using other acceptable means; if so, reference is likely to be made to the document by inspectors to demonstrate the level that has to be achieved. Equally, while it has no legal force, its standing as agreed practical guidance means that it may be referred to in a court or tribunal to demonstrate the standards that need to be met under the law.

ACKNOWLEDGEMENTS

CATIAC wishes to thank the members of the Working Group who by freely giving so much of their knowledge and experience enabled this document to be produced.

Members of Working Group

Mr A E Jones, Health and Safety Executive (*Chairman*)
Mr B Bonworth, Coats Viyella Plc
Mr P Fryer, Transport and General Workers Union
Mr J Gordon, British Textile Technology Group
Mr B G Hazel, Textile Finishers' Association
Mr J B Laird, Health and Safety Executive
Dr A P Lockett, A Lockett and Company Ltd
Mr S Mortimer, Health and Safety Executive
Mr D Palmer, Tootal Group Plc
Mr M Penson, Corah Plc
Mr G K Wilson, National Union of Hosiery and Knitwear Workers
Mr L M Wrennall, British Textile Machinery Association
Mr C H C Cottle, Health and Safety Executive (*Coordinator*)

CATIAC also wishes to thank those companies who provided valuable assistance by allowing photographs to be taken to illustrate the document.

Companies assisting with photographs

Bridge Hall Dyeing and Finishing Company, Heap Bridge, Bury
Carrington Novare, Ramsbottom, Bury
Dorma, Chinley, Stockport
P W Greenhalgh and Co Ltd, Newhey, Rochdale
Jolly and Jackson Ltd, Horwich, Bolton
The Standish Company, Worthington, Wigan
Stott and Smith Group Ltd, Horwich, Bolton
Strines Textiles Ltd, Strines, Stockport
Tonge and Young Ltd, Crumpsall, Manchester
Tootal Batik, Hyde, Cheshire

INTRODUCTION

1   The aim of this document is to identify the hazards associated with plant and machinery used in the finishing sector of the cotton and allied textiles industry, and to recommend precautions and control measures so that danger and ill health are avoided.

2   In finishing works, plant and machinery accidents continue to occur all too frequently with resultant injuries ranging from the relatively slight to those involving serious, permanent damage and even death.   Accidents associated with roller intakes and scalding or corrosive liquor are often particularly severe. Other typical accidents involve unsafe use of electricity or unsafe maintenance work.   There can also be health hazards from dust, fumes and vapour given off by process equipment.

3   Satisfactory solutions to many of these problems need not be complex, nor are they beyond current technical knowledge. What is really required is a willingness and determination by all concerned to implement the straightforward, effective control measures and precautions which are already available in most cases. Accordingly, this document offers practical health and safety guidance to those involved in the design, manufacture, installation, operation and maintenance of plant and machinery used in the preparation, colouration and finishing of cotton and allied textile materials, whether these materials are processed as loose stock, yarn, fabric or garments.

4   The full range of hazards associated with plant and machinery is considered in this document. Some emphasis is placed on mechanical hazards, but non-mechanical hazards are also assessed together with the problems which can arise during maintenance work. In any case, only rarely can safety be achieved by means of physical guards or safety devices acting alone.

Also required are safe working procedures and effective maintenance routines which in turn imply high standards of training, instruction and supervision of both operatives and maintenance staff.

5   So that these issues are fully addressed, the document is divided into:

Part 1 : General considerations;
Part 2 : Specific hazards and precautions;
Part 3 : Further information;

and to obtain maximum benefit, readers are advised to make use of all three parts. In particular, a close study of *General considerations* is advised before moving onto Part 2 to find the plant and machinery items of specific interest. Additional reference material which may be required is listed at Part 3: *Further information.* A most useful reference is BS 5304:1988 *Safety of machinery*[50] which is regarded as essential reading.

6   Finishing techniques are in a continuous state of change and development with novel processes and operations being introduced as others fall from popularity. Inevitably, the document cannot cover every conceivable hazard found or likely to be found.   If there are doubts about the safety of a particular process or plant item, advice should be obtained from the machine maker, the relevant trade association or the Health and Safety Executive.   If necessary, the matter could be referred to the Cotton and Allied Textiles Industry Advisory Committee for further consideration.

7   In some finishing works, because of local conditions or special circumstances, there may be a wish to adopt precautions and control measures other than those recommended in this document. Such initiatives are encouraged provided that at least an equivalent standard of safety and protection is achieved.

PART 1 : GENERAL CONSIDERATIONS

LEGAL REQUIREMENTS

8   The guidance contained in this document takes into account the requirements imposed by the relevant legislation which is principally the Health and Safety at Work etc Act 1974,[2] the Factories Act 1961[1] and Regulations and Orders made under those two Acts. A brief summary of the legislation follows at paragraphs 9 to13 and further information is available in references 1 to 21 listed at Part 3 of the document.

Health and Safety at Work etc Act 1974

9   This Act places a general duty on employers to ensure the health and safety at work of their employees. The duty includes the provision of safe plant and machinery, a safe working environment and all necessary information, instruction, training and supervision.

10   Employees have a duty to take reasonable care of their own safety and must co-operate with their employers on health and safety matters.

11   Plant and machinery designers, makers and suppliers have a duty to make and supply equipment which is safe to use, and must also provide adequate information on safe operation.

Factories Act 1961

12   More specific duties relating to hazards and working conditions in factories are imposed by this Act. Requirements are laid down concerning, for example, guarding of machinery; inspection and test of specified types of plant and machinery; safe means of access and safe place of work; precautions against hot or corrosive substances; control of the general working environment to include cleanliness, heating, lighting and ventilation.

Regulations and Orders

13   A number of Regulations and Orders made under the Health and Safety at Work etc Act 1974 or the Factories Act 1961are applicable to specific hazards and processes encountered in textile finishing. Most of the Regulations of more recent origin are supported by Guidance or Approved Codes of Practice which help to give a practical interpretation of the legal requirements.

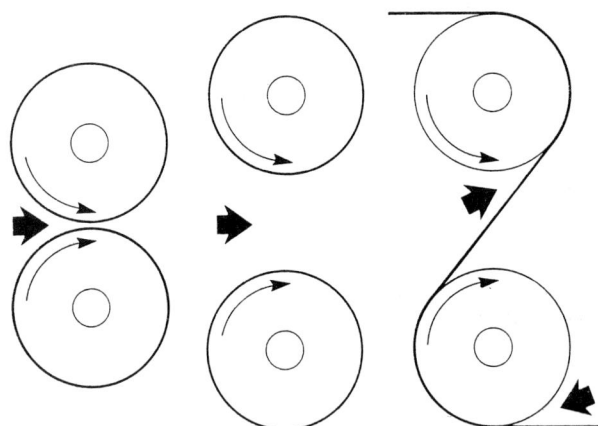

Danger	Danger	Danger
Where rollers are in contact	Where gap between rollers if less than 100 mm (4 inches)	Where blanket or fabric runs onto a roller

Fig 1 Examples of roller intakes

MECHANICAL HAZARDS

Identification of mechanical hazards

14   Examples of commonly found mechanical hazards include:

(a)   *transmission machinery* such as chain and sprocket drives, belt and pulley drives, and gears;

(b)   *rotating machinery* such as shafts, couplings and rollers;

(c)   *reciprocating machinery* which creates shear traps between moving parts and fixtures;

(d)   *machinery having considerable momentum* which takes time to come to rest when the power supply is isolated.

(e)   *intakes* between pairs of rollers or cylinders;

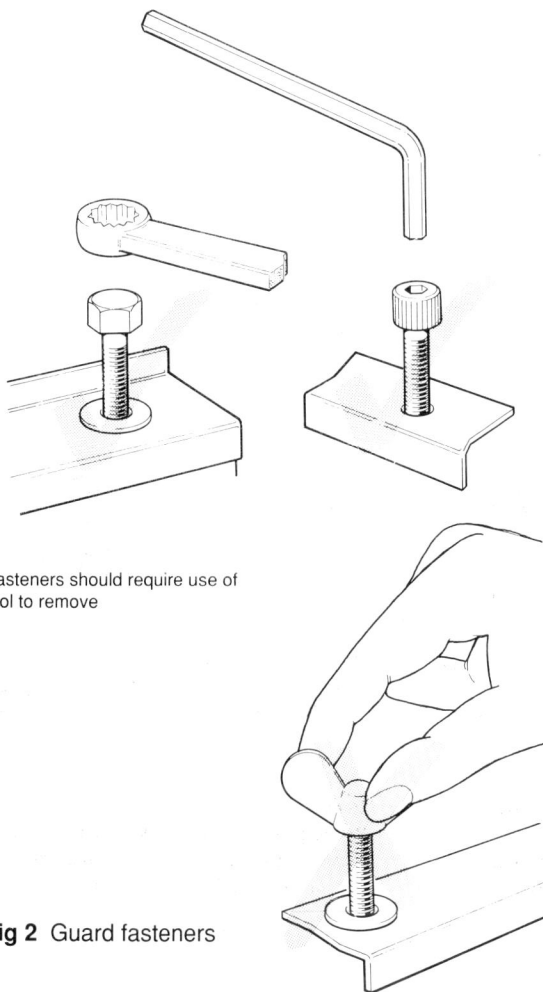

Fasteners should require use of tool to remove

Fig 2 Guard fasteners

Fig 3 Transmission drive enclosed by sheet metal guard

(f) *intakes* where fabric or a blanket runs onto a roller or cylinder.

15 **Intakes at rollers and cylinders are among the worst mechanical hazards** in textile finishing, still being responsible for numerous crippling injuries to fingers, hands and arms. **The rotating surfaces do not need to be in contact for danger to arise:** experience shows that all inrunning surfaces where the gap is 100 mm (4in) or less should be regarded as a hazard. Danger can still be present even if only one of the rollers or cylinders is power driven, with the other acting as a guide or idle roller, rotated by the motion of the fabric through the machine.

16 For each mechanical hazard where safety by design or safety by position cannot be achieved, effective guards or safety devices **must** be provided so that danger is minimised. Detailed advice is contained in BS 5304:1988 *Safety of machinery* and the reader is recommended to make full use of that publication to supplement the guidance which follows at paragraphs 17 to 29.

Guard design and construction

17 All guards provided should be of robust construction sufficient to withstand the environmental and process conditions. They should be able to tolerate normal wear and tear, foreseeable damage and chemical attack. If the guard has to be used as a working platform or means of access, it should be strong enough for those purposes.

18 For the sake of durability and simplicity, fixed guards are usually preferable to complicated safety systems. It should only be possible to remove or to open fixed guards using a tool or key such as a screwdriver, spanner or hexagonal key: wing nuts or similar easily removable fasteners are not acceptable. The use or possession of tools and keys by unauthorized persons should be prohibited.

19 Guard design should be such that it is not possible to make contact with dangerous parts by reaching through, over, under or behind the guards. Thus in some circumstances, return, back or top panels may need to be provided.

Fig 4 Aperture in mesh guard to permit machine lubrication

(b) either the act of opening the guard causes the machinery to stop or, preferably, the guard can only be opened when all dangerous movement has ceased.

22   To achieve a high degree of user acceptance, guards and safety devices should be designed in accordance with sound ergonomic principles, the aim being to cause minimal interference with production and maintenance routines. The need for lubrication of the machinery should be considered with arrangements made where possible for lubrication to be carried out with the guards in place. This may be achieved by extending lubrication pipes and grease nipples to points outside the guards or by incorporating small apertures in the guards which would admit the nozzle of an oil can or grease gun but would be too small to allow personal contact with dangerous parts. Ergonomic data including permissible openings in guards are given at Appendix A to BS 5304:1988 *Safety of machinery*.

23   Guard design should also permit safe adjustment to the machinery, with adjustments normally being made from a position outside the guard or when the machinery is at rest. If such arrangements are not feasible apertures may again be incorporated in the guards to allow the necessary hand access so long as there is no possibility of personal contact with dangerous parts.

24 **Most careful attention should be given to achieving safety at inrunning rollers and cylinders,** and the preferred solution is to provide a clear gap of at least 100 mm (4 in) between the inrunning surfaces. Where this is not possible because a lesser or zero gap is unavoidable, the traditional guarding method is to provide a fixed nip bar. Nip bars can be effective if kept in close adjustment to the rotating surfaces, with the gaps not exceeding 6mm (0.25 in), but experience shows that this vital requirement is frequently not satisfied. Thus intakes are readily created between the bar itself and the rotating surfaces giving a situation at least as dangerous as no guard at all. This is a particular problem with round nip bars which is why heavy plate or angle section bars are preferred. Also, a nip bar is not suitable where the nip position changes unless the bar is designed to self adjust accurately which is very difficult to achieve in practice.

20   Where guards or parts of guards have to be opened or removed frequently, interlocked safety devices which are reliable and which cannot be easily defeated should be incorporated. The method of interlocking is open to choice and may be mechanical, electrical, hydraulic, pneumatic or any combination of these. If electrical interlocking is used, beware of the wet and corrosive conditions in the typical finishing works which can cause reliability problems. Also, a normally open electrical limit switch is not, on its own, suitable for safety purposes since it is liable to fail in an unsafe mode and is easily defeatable. Guidance on choice of methods of interlocking is given in BS5304:1988 *Safety of machinery*.

21   Whatever interlocking system is used, it should be connected to the machinery such that:

(a)   the machinery cannot operate until the guard is closed; and

4

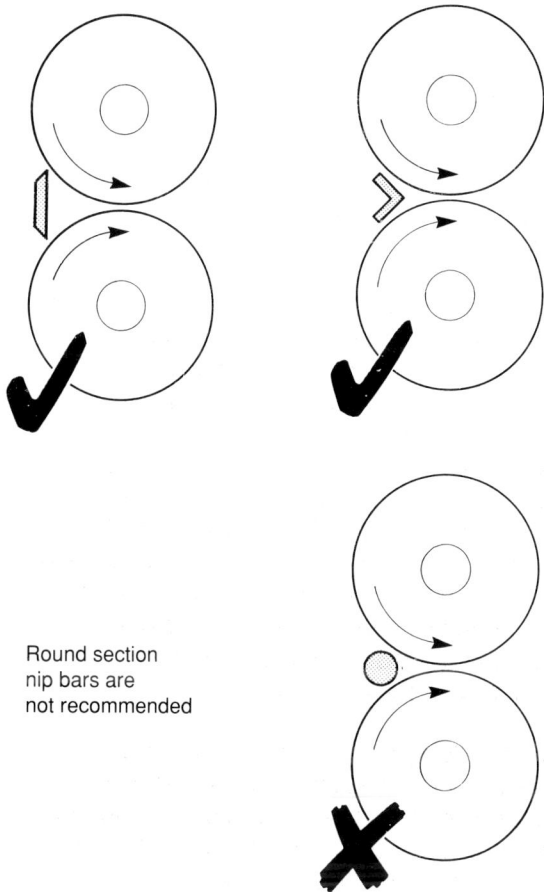

Round section
nip bars are
not recommended

Fig 5 Design of nip bars

Fig 6 Round nip bar at mangle, dangerous access
possible beneath bar

25 Because of these limitations, **it is strongly recommended that hinged, interlocked guards are provided at roller intakes** which cannot be made safe by providing an adequate clear gap, unless there is no difficulty in maintaining at all times a sturdy, close fitting nip bar or some other type of fixed guard.

Safety at unfenced machinery

26 If all the factors discussed at paragraphs 17 to 25 are taken into account when designing guards and safety devices, any necessity to operate machinery with guards removed can be avoided in the vast majority of cases. Nevertheless, within the terms of existing legislation (Sections 15 and 16 of the Factories Act 1961 and the Operations at Unfenced Machinery Regulations 1938[4]) **certain operations with guards removed are permissible, but only if:**

(a) the requirements of the Act and the Regulations are strictly observed, paying particular attention to the training and authorisation of those who are to carry out the work; *and*

(b) approach to the unguarded machinery is only for the purposes of examination, or for any lubrication or adjustment shown by that examination to be immediately necessary; *and*

(c) the examination, lubrication or adjustment can only be carried out with the guards removed and with the machinery in motion.

27 The relaxations allowed under the Unfenced Machinery Regulations 1938 should be used as a last resort only. If there is a safety problem concerning examination, lubrication or adjustment with the machinery in motion, the user should first establish whether that problem can be overcome by alterations to the machinery or its guards and safety devices.

Safety by position

28   The concept of safety by position can be an illusion and has led to many serious and fatal accidents. Moving parts which are out of reach to a person at or near floor level are safe by position while that person remains at floor level but become dangerous if approach is made for any purpose unless the plant is stopped and properly isolated. Reasons for approach could include threading-up, maintenance work or the need for access by contractors such as window cleaners or roofers. Means of access might be a ladder or steps, or by the person clambering up pipework or the machine framework.

29   The term safety by position is recognised in law but it relies upon the maintenance of a safe system of work, namely proper isolation of plant, which can fail. For this reason it is recommended that all dangerous parts, even if normally out of reach, are guarded in the usual way.

NON-MECHANICAL HAZARDS

Chemicals and hot liquor

30   Many of the process chemicals and dyestuffs used in textile finishing are hazardous to health either by skin contact, inhalation or ingestion. Hot liquor, even if non corrosive, can also be very hazardous simply by virtue of its temperature.

31   Chemical hazards can arise during normal plant operation, during plant maintenance or breakdown, from dangerous combinations of chemicals or from accidental spillage. A full and detailed appraisal of such hazards and the precautions necessary is outside the scope of this document and readers requiring further information are referred to the following publications:

(a)   *Guidelines for the safe storage and handling of non-dyestuff chemicals in textile finishing;*[42]

(b)   *Safe handling of dyestuffs in colour stores;*[53]

(c)   Approved Codes of Practice L5: *Control of substances hazardous to health and Control of carcinogenic substances.*[17]

32   In practice, many chemical and hot liquor hazards can be controlled to a significant extent by means of suitable plant and machinery design, and by following safe working procedures for normal operation, maintenance tasks and emergencies. These issues are addressed below at paragraphs 33 to 38, at *Dust and Fumes* paragraphs 39 to 41, and at *Maintenance* paragraphs 73 to 78.

33   A hazard sometimes having a fatal outcome is that of operatives falling into tanks, pits or similar open vessels containing scalding or corrosive liquor. To minimise the risk, the rims of such vessels **must** be at least 920 mm (3ft) above the adjacent floor or working platform, or **must** be fenced off to that height, or the top of the vessel **must** be securely covered over. Similarly no ladder or gangway may be placed above the vessel unless the ladder or gangway is fenced to a height of 920 mm (3ft) or the vessel is securely covered. Full legal requirements are specified at Section 18 of the Factories Act 1961.

34   The one exception to the above rules concerns vessels subject to the Kiers Regulations 1938[3]. These Regulations only require fencing of kiers and associated ladders or gangways over the kiers to a height of 840 mm (2ft 9in) rather than 920 mm (3ft). However see commentary at *Kiers* paragraphs 121 to 127 where for consistency the recommendation is to meet the 920 mm (3 ft) standard.

35   Some vessels heated by live steam tend to overflow if left unattended. The steam shut off valve should be located in a safe position well away from the vessel so that the operative can turn off the steam supply without being endangered by the boiling liquor. Automatically controlled plant which reduces the risk of overflowing can also enhance safety.

36   Chemical and liquor lines should be colour coded or otherwise marked as to their contents. See BS 1710:1975 *Specification for identification of pipelines.*[44] The functions of controls, pumps and valves should be clearly identified unless the functions are obvious.

37   Operatives and maintenance staff should be alert to the safety implications when:

(a)   operating control valves or taps in feed, circulation or drain lines;

(b)   starting or stopping pumps;

(c)   breaking into lines for repair and maintenance purposes.

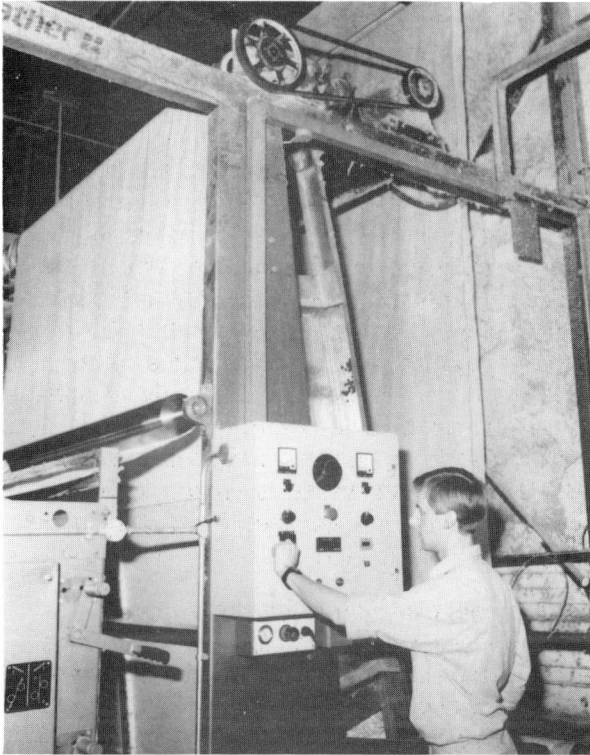

Fig 7 Belt and pulley drive, safe by position to a person at floor level only

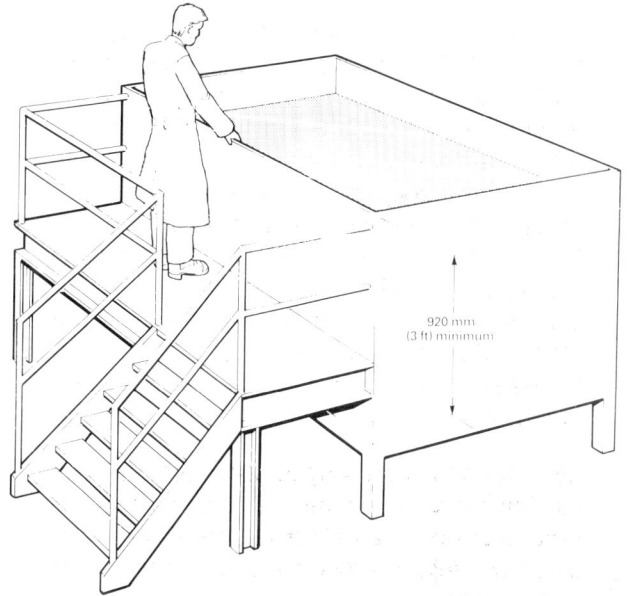

Rims of open tanks or vessels containing hot or corrosive liquor should be at least 920 mm (3 ft) above adjacent floor or working platform, or should be fenced off to that height

920 mm
(3 ft) minimum

Fig 8 Safety at open tanks or vessels

Fig 9 Fencing of open tank

Fig 10 Automatic piped dispensing of chemicals

38 Where a contact risk from chemicals or hot liquor remains, despite the use of engineering controls and other measures, suitable protective equipment should be provided. As a minimum, a face shield is necessary plus impervious apron, gauntlets and footwear. The face shield as required by the Protection of Eyes Regulations 1974[6] should conform to the chemical eye protector standard as given in BS 2092: 1967 *Specification of industrial eye protectors.*[45]

Dust and fumes

39 Dust and fumes arising from process plant should be adequately controlled so that contamination of the workroom air is kept below the relevant occupational exposure limit. See HSE Guidance Note EH40 *Occupational exposure limits,*[40] revised annually, and Approved Codes of Practice L5: *Control of substances hazardous to health and Control of carcinogenic substances.*

40 If dust or fume levels are found to be excessive, use of personal respiratory protective equipment should only be regarded as a temporary expedient until adequate control of the problem is achieved by engineering means. Possible control measures include enclosure of the dusty or fume producing process and provision of local exhaust ventilation. The exhaust ventilation system should be designed and installed by competent ventilation engineers, and requires proper maintenance to ensure efficient operation.

41 In some circumstances, process fumes extracted from the finishing works can cause environmental nuisance to the local population. Environmental nuisance is a matter for the environmental health officer of the local authority who should be consulted if necessary.

Fire and explosion

42 In processes such as singeing and cropping where dust and other combustible residues can accumulate, regular cleaning is necessary to avoid the flash fire risk. Separation of high fire risk processes from other processes is recommended and built in fire extinguishing systems may also be necessary. Advice on general fire precautions to cover fire drills, fire alarms, fire fighting equipment and means of escape should be obtained from the fire prevention department of the local fire service.

Fig 11 Explosion panels on top of stenter oven

43 Solvent evaporating ovens heated by any means and all **direct** gas fired ovens can give rise to significant fire and explosion risks. Advice on the precautions to be taken is contained in booklet HS(G)16 *Evaporating and other ovens.*[22] The main requirement is for adequate explosion relief and in the case of gas fired plant, appropriate gas controls. Technical advice on the latter item should be obtained from a qualified industrial gas engineer.

44 At ovens where there is an explosion risk, adequate explosion relief should always be provided where technically feasible, irrespective of the sophistication of the oven controls. The relief, in the form of explosion doors or lightweight explosion panels, should be sited to vent to a safe place where personnel are unlikely to be present. Relief distributed along the top of the oven is usually preferred.

45 With some types of oven, especially those used at stenters, it is possible for the internal ductwork to effectively divide the oven chamber

into upper and lower compartments. If this is the case, separate relief for both compartments should be provided.

Pressure plant

46 Pressure plant such as steam boilers, steam receivers and air receivers requires proper installation, operation and maintenance to ensure safety. The periodic thorough examinations required by law should be carried out by a competent person and many users look to their engineering insurance company for this service. In some cases non-destructive testing is an integral part of the thorough examination. For new plant, an initial examination and test may also be required.

47 The examination reports should be kept available for easy reference and a responsible person such as the works engineer should ensure that any repairs or modifications called for by the competent person are carried out. A separate log for each major plant item is advised.

48 Most other types of pressure plant including high temperature dyeing machines which contain fluids at more than 0.5 bar (7 psi) above atmospheric pressure are also pressure systems as defined in the Pressure Systems and Transportable Gas Container Regulations 1989[12], and require periodic thorough examination by a competent person. The user should liaise with the competent person to determine, for each plant item, the necessary scope and frequency of the examination routines.

49 The 1989 Regulations do not come fully into effect until 1 July 1994 but in the meantime special transitional provisions apply. Guidance is contained in the following publications.

(a) *Guide to the pressure systems and transportable gas containers regulations 1989*;[20]

(b) Approved Code of Practice COP 37: *Safety of pressure systems*.[21]

Lifting equipment

50 Lifting machinery and lifting tackle should only be used by trained personnel familiar with the risks involved and the precautions to be taken.

Safe slinging methods should be adopted and any unauthorised practices eliminated by means of effective instruction and supervision.

51 Lifting equipment needs to be suitable for the work in hand and should be properly installed and maintained to ensure safety. Particular care should be given to the maintenance of top hoist limits on electric blocks. An inoperative top limit can easily lead to overwinding and failure of the suspension rope.

52 The periodic thorough examinations required by law should be carried out by a competent person and many users look to their engineering insurance company for this service. Lifting machinery should be thoroughly examined at least once every 14 months, lifting tackle at least once every 6 months. There is also a requirement for thorough examination and test before the equipment is taken into service. The examination reports should be kept available for easy reference and a responsible person such as the works engineer should ensure that any observations and defects noted by the competent person receive attention.

Electrical equipment

53 All electrical equipment should be designed in accordance with good engineering practice, for example as recommended in BS 2771:1986 *Electrical equipment of industrial machines.*[46] The equipment and its associated electrical systems should be properly installed, operated and maintained following the recommendations of the Institution of Electrical Engineers: *Regulations for Electrical Installations*[54] (15th Edition, 1981). Legal duties are specified in the Electricity at Work Regulations 1989[10] and further information is available in booklet HS(R)25 *Memorandum of Guidance on the Electricity at Work Regulations 1989.*[18]

54 The wet and humid conditions in the typical finishing works can substantially increase the electrical risks with faults such as short circuits and arcing more likely to occur because of corrosion, ingress of water and increase in conductivity due to moisture. These electrical faults may be dangerous in themselves by creating shock or burn risks. Short circuits in control systems may also be dangerous by causing inadvertent operation of machinery.

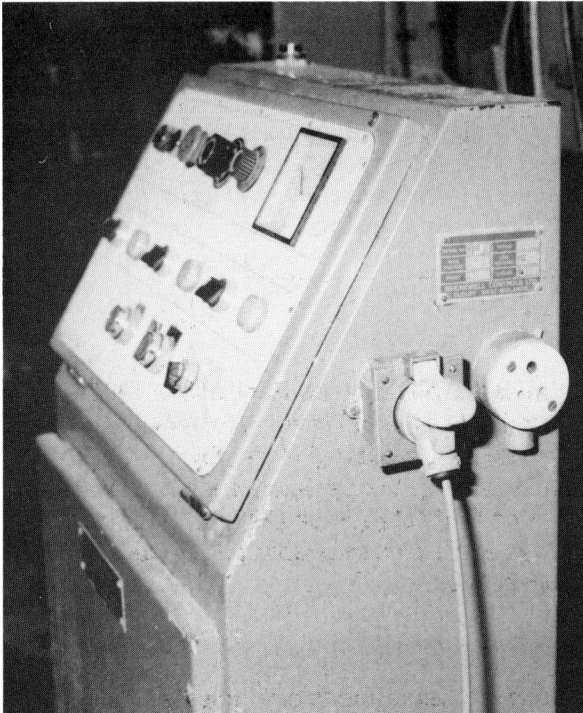

Fig 12 110 V electrical supply

Fig 13 Earth leakage circuit breaker

55   For portable equipment such as tools and sewing machines operated in wet or corrosive conditions, use of electrical power should be avoided were technically feasible, for example by using pneumatically powered apparatus.

56   In wet or corrosive conditions where use of electrical powered portable equipment cannot be avoided, the equipment should be specially constructed or protected to prevent danger. Apparatus operating at extra low voltage is particularly suitable, for example battery powered tools and handlamps. In less severe environments, reduced voltage apparatus such as 110 V centre- tapped-to-earth systems which work from portable transformers or low voltage sockets could be used.

57   Safety can also be enhanced by current operated earth leakage circuit breakers (residual current devices) set to operate at 30 milliamps leakage current within a tripping time of 30 milliseconds. However such devices can fail and should be regularly tested to ensure correct operation.

58   Conductors and their protective coverings should be properly maintained with additional protection given to cables or leads which are liable to suffer mechanical damage. Armoured cable may be needed in some applications. Care should be taken to ensure that insulation material is protected against chemical attack.

59   Further guidance is given in HSE Guidance Notes PM32 *Safe use of portable electrical apparatus*[38] and GS27 *Protection against electric shock*.[36] HSE Guidance Note GS37 *Flexible leads, plugs, sockets etc*[37] is also relevant.

60   Enclosures for fixed electrical apparatus such as switchgear and fuseboxes, unless well separated from wet areas, should be to a protection standard of at least IP55 as defined in BS 5490: 1977 *Classification of degrees of protection provided by enclosures*,[51] thereby giving protection against water jets. Electric motors should also be to a similar standard where necessary.

Microprocessor controlled plant

61   Safety and reliability problems can be introduced at microprocessor controlled plant if the safety systems are connected through the microprocessor. To avoid such problems, the safety systems should be hard wired in the conventional way and kept separate from the electronic control circuitry. Possible hazards arising

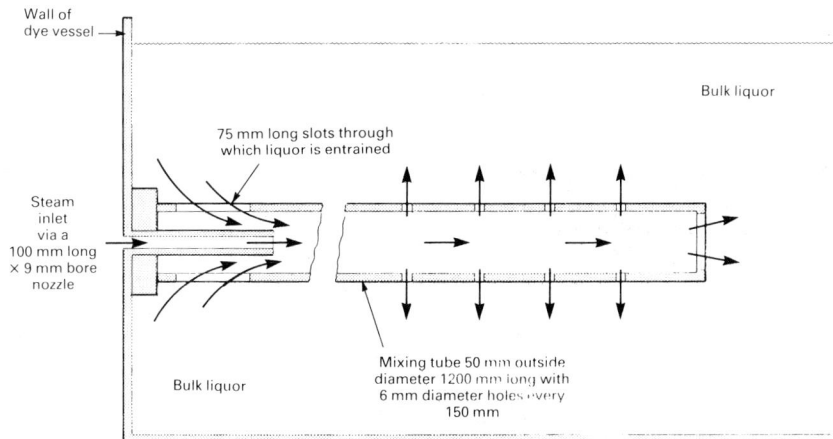

Fig 14 Control of steam noise by use of modified steam injector

from abnormal plant behaviour should be considered and maintenance risks carefully assessed.

62   Detailed advice is given in a pair of HSE publications entitled *Programmable electronic systems in safety related applications: Part 1 An introductory guide;*[34] *Part 2 General technical guidelines.*[35]

Static electricity

63   Static electricity can build up on non-conductive process materials as a result of:

(a)   friction;
(b)   contact or separation of surfaces, for example fabric passing over a roller;
(c)   cutting;
(d)   rapid temperature change.

The discharging spark can give rise to a fire risk or cause unpleasant nuisance shocks to personnel.

64   Precautions should be taken to control the static risk by providing suitable earth pathways from the plant or by installing static eliminators in appropriate positions. Static eliminators currently fitted are usually high voltage units where the current is limited to a very low, safe level. They do not incorporate radio-active sources but rely instead on a high voltage corona to ionise the air and dissipate the charge.

Ionising radiation

65   Fabric yield gauges, used for monitoring the weight or density of fabric, are sealed radio active sources, usually Beta emitters. Detailed advice on safe storage and use should be obtained from the suppliers. The equipment is not repairable on site and in the event of a fault it should be returned to the supplier or other competent repairer.

66   The Ionising Radiations Regulations 1985[8] apply and include the requirement for leakage testing to be carried out periodically by a competent person. Further information is available in Approved Code of Practice COP 16: *Protection of persons against ionising radiation arising from any work activity.*[16]

Noise

67   Prolonged exposure to high noise levels is known to result in irreversible damage to hearing. Noise also interferes with the ability to concentrate and communicate and can be assumed to be a contributory factor in some accidents.

68   Steps should be taken to ensure that personal noise exposure averaged over the working day does not exceed 90dB(A) and preferably does not exceed 85 dB(A). Measures available include:

(a)   enclosure of the noise source;
(b)   use of noise reduced machine components instead of metallic components;
(c)   proper maintenance and lubrication of machinery.

Wearing of personal hearing protection should only be regarded as a short term expedient until control of high noise levels is achieved by engineering means.

69   Legal duties are specified in the Noise at Work Regulations1989[11] and further information is available in booklet *Noise at work: Noise guides No 1 and No 2.*[19]

70   There is a significant noise problem at vessels such as jigs and washers where heating of the liquor is by the direct injection of live steam. Noise levels of up to 100 dB(A), caused by the collapsing steam bubbles, are generated for short

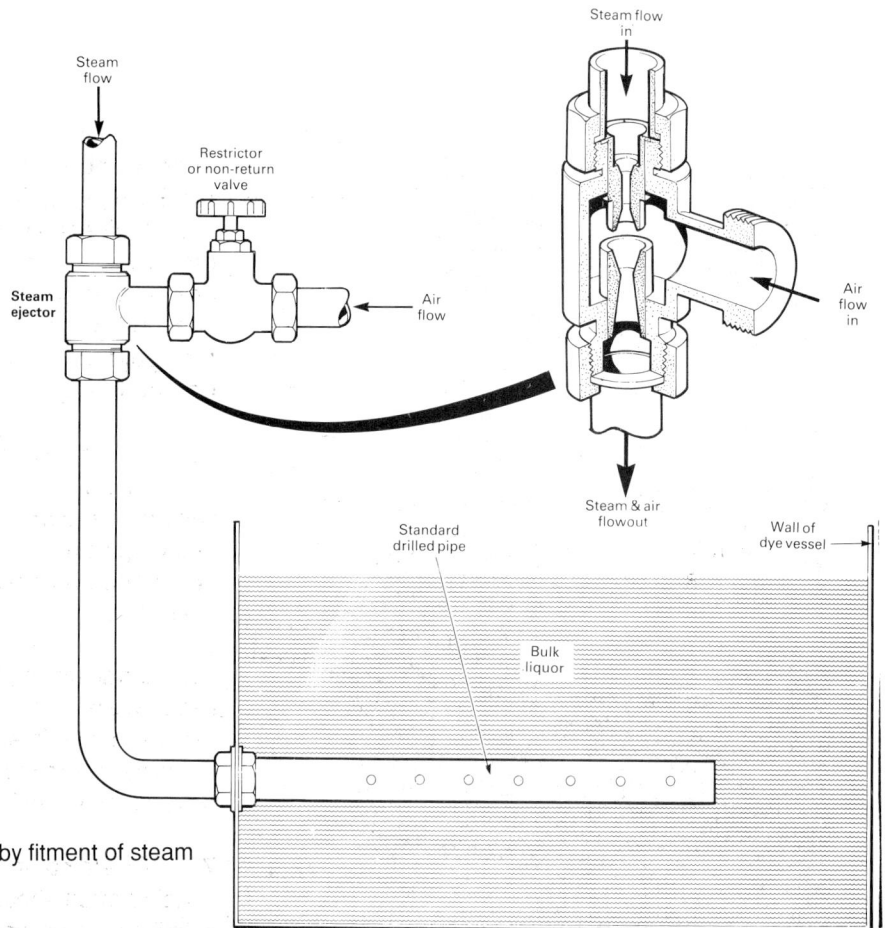

Fig 15 Control of steam noise by fitment of steam ejector

periods during the boiling up phase. These noise levels can be reduced to about 85 dB(A) or less by:

(a) modifying the steam injector (sparge pipe) to entrain liquor with the flow of steam; or

(b) fitting a steam ejector to entrain air at atmospheric pressure with the flow of steam.

71   The liquor entrainment method has proved particularly successful at a number of finishing works. Use of preheated water can also help to reduce the duration of exposure to high noise levels.

72   Transmission drives are sometimes noisy, but in most cases the guards for protection against the mechanical hazards can also be designed to act as efficient acoustic enclosures where necessary.

MAINTENANCE

73   Regular inspection and maintenance by competent staff is essential if plant and machinery is to operate efficiently and safely. Problems frequently arise when guards and safety systems originally of sound design and construction fall into disrepair or fail to function correctly because proper maintenance has been neglected. In the typical finishing works, the wet and corrosive conditions and heavy wear and tear combine to

Fig 16 Safe access to top of boiler

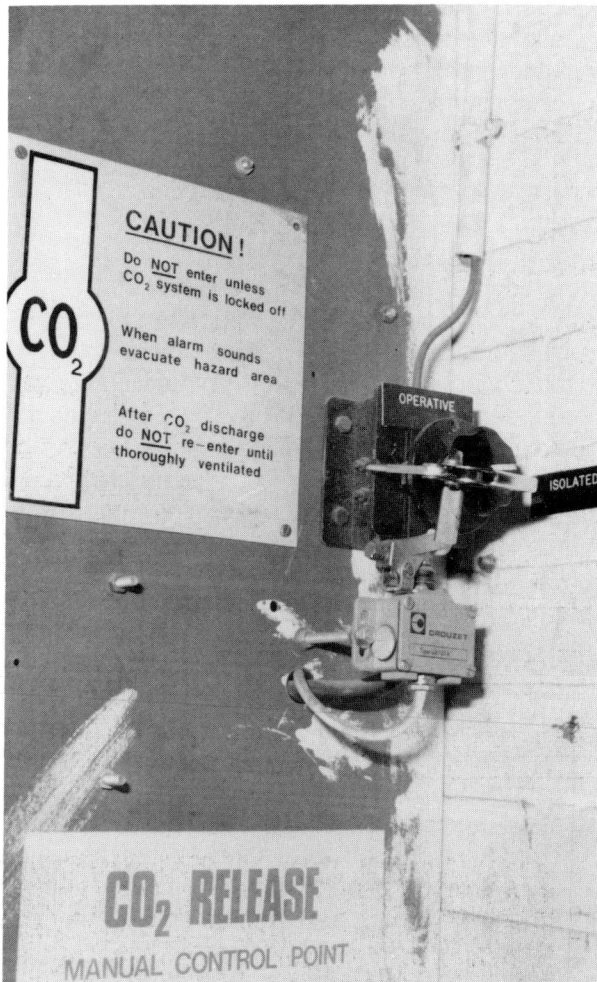

Fig 17 Lock-off procedure for carbon dioxide fire extinguishing system

increase the maintenance load. Where necessary, in house inspection and maintenance staff should be supplemented by outside contractors and service engineers.

74   Maintenance work is itself potentially dangerous and the risks involved are often quite different from those facing production operatives.

There is the chance of inadvertent starting of machinery, possibly by a second person, and risks are introduced when guards need to be temporarily removed. Consideration should also be given to problems such as working at height, entry into confined spaces and cutting or repair work using the application of heat. Exposure to chemicals and hot liquor may also give rise to significant risks.

75   Safe systems of work for maintenance tasks should be drawn up and enforced. Arrangements should be made for the use of protective clothing and respiratory protective equipment as necessary, and emergency procedures should be clearly specified.

76   Cleaning of machinery introduces its own hazards and should normally be carried out while the machinery is at rest. Cleaning while the machinery is in motion is only acceptable if it can be accomplished in complete safety.

77   Where contractors are employed, special consideration should be given to the risks their activities may cause, not only to themselves but also to the regular, full time work force. Likely problems such as means of access, proper isolation of plant and avoidance of chemical and hot liquor hazards should be anticipated and overcome rather than left to chance. This implies close liaison between management at the finishing works and the contractors.

78   The scenarios of numerous fatal accidents associated with maintenance work, and the lessons to be learnt, are given in a three part series of HSE booklets entitled *Deadly maintenance: a study of fatal accidents at work.* The three booklets cover *Plant and machinery,*[31] *General hazards*[32] and *Roofs.*[33]

PART 2 : SPECIFIC HAZARDS AND PRECAUTIONS

EXPLANATORY NOTE

79 Any classification of plant and machinery used for the preparation, colouration and finishing of cotton and allied textile materials is complicated by the numerous different ways in which the materials can be offered for processing, whether as:

(a) *loose stock* - staple or tow;

(b) *yarn* - hanks, packages, ropes or beams;

(c) *fabric* - woven, non-woven, knitted or narrow; treated in open width or rope form; continuously processed or batchwise;

(d) *garments* - hosiery or other knitted goods; complete garments or garment pieces.

80 There is also a wide variety of process options that a material may or may not undergo. For example in the case of a woven fabric, it may be dyed and printed, dyed or printed or not coloured at all. Also, a particular type of machine or plant item is not necessarily confined to an individual sub-process of the overall finishing treatment. For example, mangles, drying cylinders and fabric take-off units can all be encountered in either preparation, colouration or finishing operations.

81 To minimise these complications, no classification by type of material processed has been attempted in this Part 2 of the document. Instead, the plant and machinery has been grouped as appropriate under either preparation, dyeing, printing, finishing or inspection and making-up processes. Items which could be encountered more than once have generally been allocated to preparation processes, the only exception being drying plant which has been brought together under finishing processes. In case the reader has difficulty in locating a particular item, an alphabetical index has been provided at paragraph 280, cross referencing each item by its paragraph number.

82 The plant and machinery described is full scale equipment. If laboratory scale equipment is used, for instance in test and quality control

departments, comparable safety standards are needed because even at the laboratory scale serious incidents have occurred, especially accidents associated with roller intakes, hot liquor and electricity.

PREPARATION PROCESSES

Sewing

83 Heavy duty mobile sewing machines are used for joining fabric pieces for ease of subsequent processing. Some machines are hand operated but most are powered by mains electricity and the electrical risks can be serious, bearing in mind the wet and corrosive conditions which are likely to prevail in the typical finishing works. For this reason pneumatically powered sewing machines are strongly recommended, certainly as new or replacement equipment.

84 Where the use of electrical power continues for the time being, a high standard of electrical maintenance is essential, both of the machine itself and its wiring. Regular visual inspections should be carried out, particularly for damaged plugs and cables, and periodic earth continuity checks are necessary. Trailing cables should be kept as short as possible and should not be run over by trucks or other heavy items. Use of armoured cable may be appropriate in some circumstances.

85 In addition to the precautions described at paragraph 84, it is recommended that the power supply is at 110 V through transformers centre tapped to earth. Use of earth leakage circuit breakers (residual current devices) set to operate within 30 milliseconds at 30 milliamp fault current can also enhance safety. However, such devices can fail and need periodic testing, a test button being provided for this purpose.

86 Plugs and sockets at sewing machines are liable to heavy wear and tear so industrial grade connectors are recommended, conforming to BS 4343:1968 *Specification for industrial plugs, socket outlets and couplers for AC and DC supplies.*[49]

Brushing

87 Brushing machines are used in the preparation line and prior to printing for the beating and brushing of fabric to remove loose dirt and fibre. The machines incorporate rotating bristles

Fig 18 Exhaust ventilation at brushing machine

Fig 19 Interlocked guards at cropping machine, exhaust trunking also shown

and are different from raising machines encountered as a finishing process. The brushing operation may be combined with cropping.

88   Size dust and other particles tend to accumulate inside the machine even though exhaust ventilation is applied. A fire risk is created so access panels are provided for cleaning purposes. To prevent contact with internal moving parts, principally the rotating beaters or brushes, the access panels should be effectively interlocked through the power supply so that the plant can only be set in motion when all the panels are fully closed. Opening of any panel should cause the plant to stop.

Cropping

89   Cropping or shearing machines produce a smooth and level fabric surface by passing the fabric under a set of rotating, spiral cutters. There are two types of machine, each working on a principle similar to that of a cylinder lawnmower. In the preparation line, the machine is used to take off loose woven threads from the fabric while as a finishing process, the machine is rather more accurate, for precision work such as the creation of velour or velvet type finishes.

90   In both cases the spiral cutters are extremely dangerous and all access should be prevented until they are at rest. It may be sufficient to provide a fixed enclosure with a letterbox type of opening permitting entry of the fabric but not the fingers, but where frequent access is necessary, interlocked covers should be provided. The interlock should ensure that:

(a)   the cutters cannot be set in motion until the covers are closed; and

(b)   the covers cannot be opened until the cutters have come to rest.

A time delay or rotation sensing device is necessary to satisfy this second condition.

91   Modern machines detect seams in the fabric and automatically raise the cutters to clear the seam. The interlock arrangement should take this factor into account, with the guards remaining effective whenever the cutters are in motion including when they are raised.

92   The cutters are only removed for regrinding in the event of major damage. Usually they are sharpened or lapped in situ using grinding paste while being driven in the reverse direction at

15

cropping speed. Trained, experienced personnel should be employed on this work, with purpose designed temporary guards in position to prevent access to the cutters. Slots may be incorporated in the guards to allow application of grinding paste as long as fingers cannot reach through. Care should be taken since the run-down time of the machine will be extended because of no load on the ledger blade.

93   At the rear of some machines, there is a dangerous intake between the take-off roller and tension roller and it may also be possible to reach the cutters from the rear. New machines tend to be fully enclosed at this point but additional fixed guarding may be required on older plant. Where the take- off roller is clothed with card wire, full enclosure to prevent entanglement with the roller is needed even if there is no intake as such. High speed scroll rollers can also be dangerous, particularly twin scroll rollers which should be guarded.

94   There is a considerable fire risk associated with the flock dust and oil arising from the cropping process. Proper maintenance of the dust collection system is important.

Slitting

95   As a preparation process, slitters are used for the opening up of tubular knitted fabric and may be combined with plaiting. As a finishing process, they may be fitted on stenters as selvedge cutters. Another type of unit is used for the cutting of rolls of fabric into tapes.

96   The slitting effect is achieved either by means of a high speed circular knife or by the shearing action of a pair of knives. In both cases, the knives should be enclosed as much as possible consistent with being able to use the machine. Where there are built in sharpening devices used by the machine operative, eye protection should be worn if the guard enclosure is not in itself sufficient to prevent risk from flying particles.

97   A high speed knife can sometimes appear motionless, particularly under fluorescent lighting, and visual indication of rotation is therefore advised, for example a painted mark on the knife or its spindle. Braked drive motors are also recommended to prevent excessive run-down time.

98   Where it is necessary to apply a cutting lubricant to the knife, arrangements should be made to do this safely, for example using an aerosol spray.

Singeing

99   Singeing removes surface fibres from fabric and is usually carried out before bleaching and dyeing, often in conjunction with desizing. The singeing effect is achieved by passing the fabric over a set of gas burners or close to a number of radiant panels, usually heated metal or ceramic plates.

100 Where singeing is by means of gas burners, the gas supply should incorporate a solenoid valve to cut off the fuel automatically if the plant stops, so as to prevent ignition of the fabric. Also, compressed air may need to be blown in to quench the flame and provide a rapid cooling effect. To reduce the risk of burns, the flame width should be kept adjusted so that it is only just wider than the fabric.

101 Where singeing is by means of radiant panels, the gas or other fuel supply should again be cut off automatically if the plant stops. If necessary, shutters should drop down to cover the panels or the panels should swing away to prevent ignition of the fabric.

102 There is a considerable fire risk associated with the singed particles given off. These particles tend to collect in the exhaust trunking which should be cleaned out at regular intervals. With knitted fabric, unless processed in tubular form, the edge may curl over, contacting the flame or radiant panels and catching fire. A built in fire extinguishing system is particularly important where knitted fabrics are processed.

Desizing

103 Desizing is a fabric preparation process to remove the sizes which would have been added to the warp prior to weaving of the fabric. The process may involve a padding operation or the use of a jig or some other dyeing machine.

104 Where desizing is combined with singeing, the mangle which transports the fabric from the singeing plant should be guarded in the usual way. See *Mangles* paragraphs 109 to 114.

Fig 20 Switch for steam drench fire extinguishing system at singeing machine

Fig 21 Mesh guard at stirrer

105 Hazards arising from the process chemicals should be considered, especially the potential dust problem if powdered enzymes are used.

Liquor preparation

106 Liquor and process chemicals are usually prepared in mixing vessels where they are agitated by a motor driven stirrer. The stirrer may be suspended and counterbalanced, or clamped to the rim of the mixing vessel.

107 The stirrer blades and the lower end of the drive shaft on which they are mounted can normally be regarded as safe by position. However, the upper end of the drive shaft is readily accessible where it couples to the reduction gearbox. A particular danger at the coupling is grubscrews or other projections which should be sunk to avoid risk of entanglement. Stainless steel mesh guards have also proved successful at this point.

108 High speed stirring of the contents of the vessel can cause splashing or the creation of an aerosol mist and should be avoided where

possible. In many cases mixing with the stirrer operating at a slow speed is feasible but if there is still a significant health hazard, enclosure of the top of the vessel together with installation of local exhaust ventilation is warranted. If flammable materials are processed, effective earthing is needed to prevent build up of static electricity.

Mangles

109 Mangles are used for:

(a) the partial removal of moisture from fabric;

(b) the transport of fabric through a machine or process;

(c) the impregnation of fabric with dye liquor or other chemicals (this operation being known as pad dyeing or padding).

110 In each case, the moisture removal, transport or padding effect is achieved by a pair of pressure rollers or bowls drawing and squeezing the fabric through the mangle nip. Mangles can either be horizontal, vertical or inclined. In dry finishing

Fig 22 Angle section nip bar at mangle

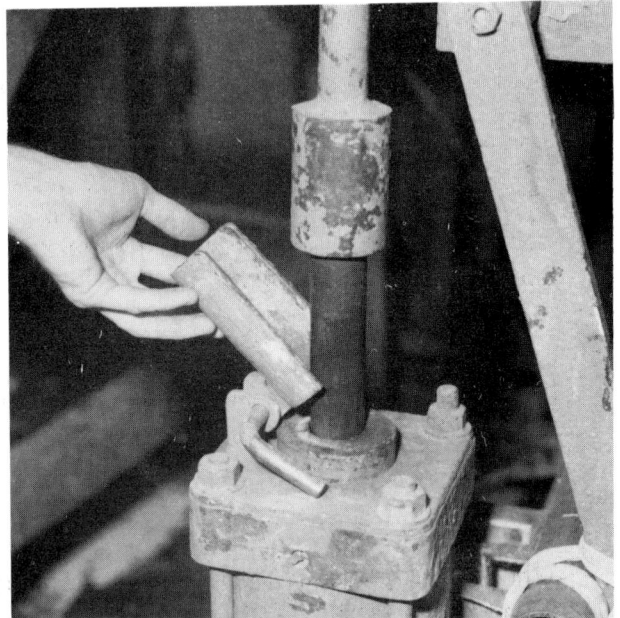

Fig 23 Stop for holding mangle bowls apart

processes, multi-bowl machines are referred to as calenders. The rollers are usually steel or steel with a composition covering but other materials are occasionally used.

111 Access to the very dangerous mangle nips **must** be prevented and it has been custom and practice in the industry to use fixed nip bars, but with limited success. **Nip bars can be effective but only if**:

(a) the bar has a cross-sectional shape which permits setting very close to the roller surfaces (round bars are less effective than plate or angle section bars); *and*

(b) the bar is kept closely set to the roller surfaces, with the gaps not exceeding 6 mm (this implies accurate positioning of the bar, avoiding the use of adjustment slots); *and*

(c) the bar is repaired or replaced if damaged.

112 Experience shows that these conditions, essential for safety, can be very difficult to achieve in practice as proved by a continuing history of serious mangle nip accidents. Indeed, **a nip bar which is out of adjustment or damaged is at least as dangerous as no guard at all** since not only is the mangle nip exposed but intakes are also created between the roller surfaces and the bar itself.

113 To overcome these problems, users are strongly advised to fit a hinged, interlocked guard where possible unless there is no difficulty in maintaining at all times a sturdy, effective and close fitting bar. An advantage of an interlocked guard is that with the power off and the mangle bowls stationary, clear access is provided for threading up purposes.

114 A nip bar does not eliminate the other hazard at threading up whereby the operative or a second person having moved the mangle bowls apart using the pneumatic control could close them again while fingers are in the gap between the bowls. Effective training and a safe system of work is required to avoid this risk, the system of work to include the use of a device to hold the bowls apart. The device could either be a manually engaged stop or an electro-pneumatic control incorporated with the interlocked guard, where such a guard is fitted.

115 Depending on the wet process immediately prior to the mangle, there may be a chemical splashing hazard.

Washers

116 Washers operate in the preparation line and after printing or dyeing, working on fabric in either open width or rope form. The term washer includes traditional types of machine known as impregnators and saturators.

18

117 The main risks are intakes at rollers and mangles, and there are also chemical and hot liquor hazards. High noise levels are likely to arise for short periods from heating by the direct injection of live steam. Modifications should be made to the steam injection systems so that the noise level in the work area does not exceed 90dB(A), and preferably does not exceed 85dB(A). See *Noise* paragraphs 68 to 71.

118 A variety of machinery including paddle machines, rotary drum washers and tumblers is used for washing and other preparation of hosiery and knitwear. See *Garment dyeing* paragraphs 168 to 172.

Pits

119 Pits are dwell or reaction chambers carrying out a function similar to J boxes except that the liquor is normally cold. They are tile lined containers up to about 5m (16 ft) deep, either set into the ground or positioned above floor level. Fabric is plaited into the pit, reacted over a period of time and then drawn out again.

120 The main risk is that of falling in so fencing around the edge at least 920mm (3ft) high should be provided. The floor of the pit has an open, ceramic structure and the risk of tripping or falling when cleaning should be considered. There may also be danger from the moving parts of the traversing piler or plaiter.

Kiers

121 Kiers are defined as fixed vessels used for boiling textile material, where the boiling liquid is circulated by means of steam or mechanical power through a pipe, channel or duct, so constructed and arranged that the liquid is discharged over the textile material and percolates through it. The Kiers Regulations 1938 (as metricated in 1981) refer. Kiers are tending to be replaced by continuous preparation ranges, but both pressure and open or atmospheric kiers are still found in use for the scouring and bleaching of fabric. The main hazards are that of scalding by hot liquor or steam when a person is inside the kier, and the danger of falling in.

122 The Regulations were brought into force following a long history of very serious scalding accidents including a quadruple fatality. Typical

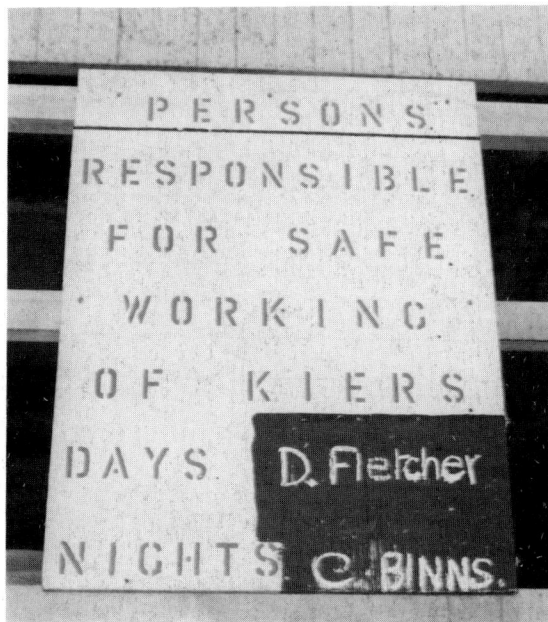

Fig 24 Safe working at kiers

circumstances would be the accidental admission or circulation of hot liquor or steam while personnel were inside the kier. To prevent such accidents, a number of practical requirements were introduced and these have stood the test of time. These requirements include:

(a) the admission of steam to be controlled by means of a screwdown valve, not merely by a tap or cock;

(b) each kier to have an individual discharge pipe, discharging to a safe place;

(c) hot liquor to be prepared in a vessel separate from the kier;

(d) an authorised person or persons to be appointed to supervise the working of each set of kiers.

123 Kiers used to be entered on a regular basis for plaiting down and similar purposes although this should not now be necessary because of mechanical piling of the fabric. In any case, regular entry to a kier is only permitted where:

(a) there is no hot liquor or hot water in that kier, hot being defined as above 40°C (105°F);

(b) the steam supply pipe has been disconnected, or the steam valve closed and locked;

(c) the liquor supply pipe has been disconnected (and the water supply pipe likewise disconnected, unless it is fitted with a non-return valve) so that the kier is isolated from all other kiers;

(d) the external circulation pipe has been disconnected, or in the case of an internal circulation pipe or puffer, the steam pipe has been disconnected or the top of the puffer blanked off;

(e) the circulation pump has been secured against accidental starting;

(f) the authorised person has displayed a notice allowing entry to the particular kier.

124 Patented disconnection devices are available to allow convenient and obvious disconnections to be made of the liquor, steam and circulation pipes.

125 If entry to a kier is required just occasionally, for sheeting up or for rectifying a fault in the fabric, this is acceptable on the authority of the authorised person if conditions (a) and (e) at paragraph 123 are complied with and if all valves and taps which control the admission of steam, liquor and hot water are closed.

126 Restrictions are placed on the dimensions and spacing of kiers to reduce the risk of falling in. Kiers should be at least 460 mm (18 in) apart and the Regulations call for the top of open or atmospheric kiers to be at least 840 mm (2ft 9in) above the adjacent platform or gangway, or the kier to be fenced off to that height. This 840 mm (2ft 9in) standard was specified at a time when it was customary for young boys to work at kiers and it was felt that any higher standard would merely encourage the boys to use say a stool or block of wood to facilitate access. Since children are not allowed to work in factories these days, the consideration is no longer applicable. It is recommended therefore that the minimum height of open or atmospheric kiers is increased to 920 mm (3ft) to equal the standard of protection required at all other open vessels containing hot or corrosive liquor.

127 A further precaution against falling in is that nobody should be allowed to sit or stand on the edge of an open kier or on the barrier fencing around it.

Fig 25 Working parts of scutcher, require guarding if accessible

J boxes

128 J boxes, named after their characteristic shape, are dwell or reaction chambers which accumulate and delay impregnated fabric as it passes through a deep trough. Both open width and rope form units are used and the boxes may either be fully enclosed at the top or open.

129 The impregnated fabric presents a chemical hazard and may also be very hot as a result of steam heating applied within the box, so in the case of open topped units, barriers at least 920 mm (3ft) high are required to prevent the risk of falling in. Access should also be prevented to the nip rollers which transport the fabric through the box, and to the oscillating plaiter. Although the plaiter zone is an unpleasant, hostile environment tending to keep operatives away, protection is still needed at this point.

130 Cleaning of the interior of the J box is a confined space problem requiring a safe system of work including use of safe means of access. Entry should be restricted until:

(a) liquor and steam valves have been securely locked off;

(b) the temperature has reduced to a safe level;

Fig 26 Door interlocks at pressure steam chamber

time for fabric processed on open width scouring and bleaching ranges. The units are either fixed in position or mobile, the latter being known as caravans. Within the chamber is a roller bed and a system of fabric draw rollers. There is usually a viewing window and internal lighting so that passage of the fabric can be observed.

134 Two pneumatically inflated seals are located at the top of the chamber permitting entry and withdrawal of the fabric. If a seal becomes damaged, perhaps by a bad sewing, the loss in air pressure should cut off the steam supply and open an exhaust valve to minimise the escape of steam through the damaged seal. Nobody should be allowed near the top of the chamber when it is under pressure in case a seal does fail.

135 Pressure steam chambers are subject to periodic thorough examination by a competent person. Close attention should be given to maintenance of the door locking devices which should incorporate interlocks to prevent the door being opened until the steam supply has been isolated and both pressure and temperature have reduced to safe levels.

Atmosphere steamers

136 Atmosphere steamers can be grouped into three main categories:

(a) roller bed steamers which are used on scouring and bleaching ranges to give the fabric a dwell period in a relaxed condition;

(b) high temperature festoon steamers which are used for dye fixation and which incorporate a long series of upper and lower rollers to transport the fabric through the plant;

(c) flash agers which are smaller high temperature units for rapid fixation of certain types of dye.

137 These steamers use either saturated or superheated steam but at nominal atmospheric pressure so they are not steam receivers and there is no pressure risk as such. The main risk is scalding, particularly at units which are open at the bottom and to which access is possible. Training of personnel is important so that entry is not made until the steam supply has been locked off and the temperature has reduced to a safe level.

(c) ventilation has been provided as necessary to clear residual fumes.

Scutchers

131 Scutchers are used to open out fabric from rope form to full width. They tend to be located high in the bleachcroft and normally there is no need for approach to the working parts while the machine is running. Under these circumstances, the working parts might be regarded as safe by position, but see comments on the concept of safety by position at paragraphs 28 and 29. Sometimes the working parts are accessible from gangways and platforms and in such cases guarding should be provided in the usual way.

132 Scutchers normally incorporate a detwisting device which may control an automatic platform on which the truck of fabric is carried. There is a risk of being hit by the moving truck as the detwisting device causes the platform to rotate. Warning notices are recommended so that personnel are aware of this risk.

Pressure steam chambers

133 Pressure steam chambers or autoclaves are closed vessels used for shortening the reaction

Continuous scouring and bleaching

138 The scouring and bleaching of fabric is frequently carried out on continuous preparation ranges which permit, without interruption, the processing of very large batches of material. The treatment can be in open width or rope form.

139 The major plant items likely to be found on continuous preparation ranges are:

(a) mangles;
(b) washers;
(c) J boxes;
(d) steam chambers;
(e) steamers;
(f) drying cylinders;
(g) fabric take-off units.

140 The size of the plant may cause communication difficulties between operatives. This problem should be anticipated and systems of work adopted to minimise danger. These systems of work should include the use of audible or visual warnings when the plant is about to be started, and a tannoy system may be justified in some cases. Suitably placed stop buttons should also be provided to protect operatives when away from the main control panel.

Mercerizing

141 Mercerizing entails the impregnation of cotton yarn or fabric with a very strong caustic soda solution which swells and plasticizes the individual fibres releasing many internal strains. As the cotton fibres swell, they tend to contract in length but this tendency is mechanically resisted by applying tension during dilution of the caustic soda at the subsequent washing stage.

142 Yarn is processed at mercerisers where tension is usually applied hydraulically. Operatives should be aware of the intake hazard between the yarn and the tensioning rollers.

143 For the processing of fabrics, both chain and chainless mercerisers are used. Chain mercerisers provide good control of warp and weft tensions but are suitable only for woven materials. Guards should be provided where the chains run onto the main drive sprockets. Chainless mercerisers, consisting of a series of tanks together with guide

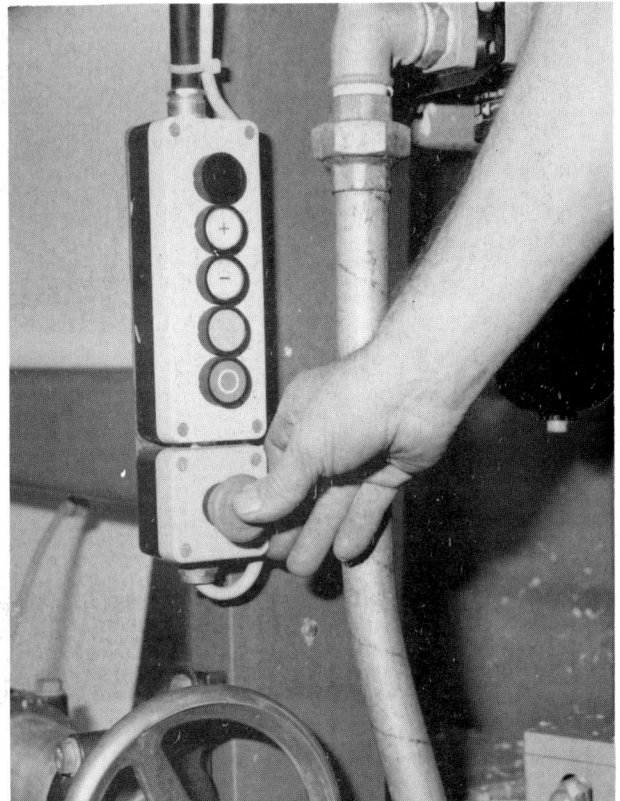

Fig 27 Stop button at continuous preparation range

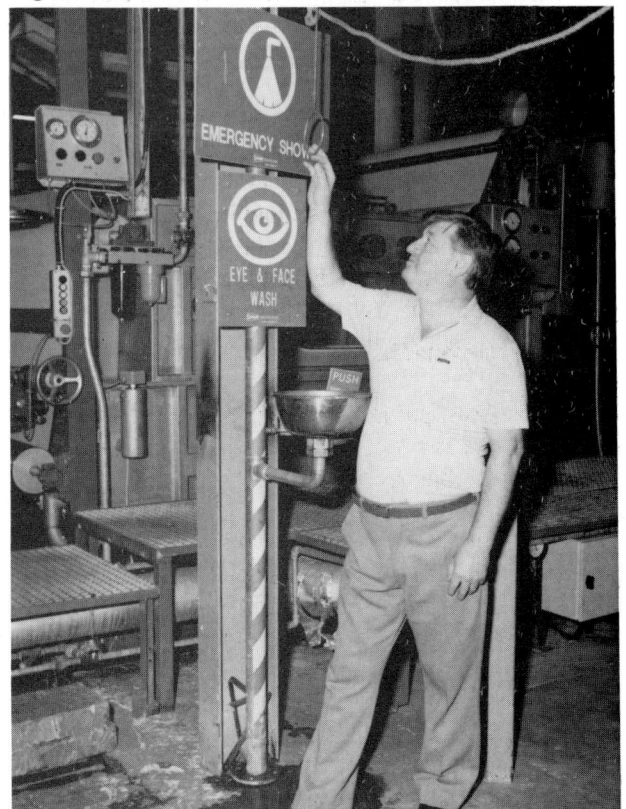

Fig 28 Drench shower and eye wash station

Fig 29 Feed rollers at plaiting down unit

rollers and mangles, give less weft ways control but with care certain knitted fabrics including tubular knits can be processed. The mangle nips should be guarded in the usual way.

144   At all variants of the mercerizing process, the caustic soda solution creates a risk of severe burns, especially to the eyes. Eye wash bottles and drench showers should be readily available. A high standard of discipline in the wearing of personal protection is also necessary.

Fabric take-off

145   Downsteam of each process the fabric being taken off is either plaited down onto a pallet or truck, or is wound into a reel or batch.

146   Plaiting down is the traditional take-off method and permits transfer of the fabric to the next process without the use of a fabric accumulator. The plaiting down unit is mounted at or above head height and comprises a pair of power driven feed rollers plus a swinging arm driven by a crank. The fabric, delivered to the swinging arm by the feed rollers, descends in a waving motion onto the pallet or truck below.

147   Dangerous parts of the plaiting down unit are:

(a)   the intake at the feed rollers;
(b)   belts and pulleys;
(c)   shear traps associated with the crank which drives the swinging arm. There may also be a risk of being struck by the swinging arm.

148   Sometimes the plaiting down unit is not readily accessible from floor level and in such cases might be regarded as safe by position, but see comments on the concept of safety by position at paragraphs 28 and 29. In any case, access will be required from time to time for threading up and maintenance purposes hence it is recommended that the working parts, even if normally out of reach, are guarded in the usual way.

149   The alternative take-off method is to use A-frame batching units to wind the fabric into reels, typically up to about 2 m (6ft 6in) in diameter containing several thousand metres of fabric. The batch may be centre-wound but more usually the drive is tangential at the surface of the batch from a pressure roller. This tangential drive introduces an intake between the pressure roller and the surface of the batch and a number of serious accidents have occurred at this point.

Fig 30 Enclosure with interlocked access gate at A-frame batcher

Fig 31 Multi-cell photelectric guard at A-frame batcher

150 Any guarding solution is complicated by the following practical considerations:

(a) for a given batch, the position of the intake changes as the batch gradually builds up in size;

(b) the A-frames on which the batches are carried are not made to standard dimensions so the position of the intake varies from one A-frame to the next;

(c) the position of the intake depends on the precise location of the A-frame relative to the batching roller.

151 These considerations imply that any guarding system, to be effective, should either be tolerant of variations in the position of the intake or should be adjustable, preferably self adjustable.

152 Taking these factors into account, the following guarding methods are acceptable:

(a) an enclosure at least 2 m (6 ft 6 in) high around the complete batch, provided with an access gate interlocked such that the batch

can only run when the gate is closed. This guarding method is effective but is only suitable where the operative does not need to approach the batch while it is running.

(b) a multi-cell photoelectric guard scanning the danger zone such that interruption of the guard screen causes the pressure roller to stop and the batching arm to lift before injury results. A degree of self adjustment for the framework carrying the photoelectric system is desirable. The system should be reliable and should not be capable of being easily defeated.

(c) a trip guard, comprising a self adjusting portion plus a hinged trip plate or flap incorporating a limit switch. Contact with the plate or flap should operate the limit switch causing the pressure roller to stop and the batching arm to lift before injury results.

153 Some users have experimented with other guarding systems but only with limited success. It needs to be stressed that the following ideas are **not** acceptable, for the reasons stated:

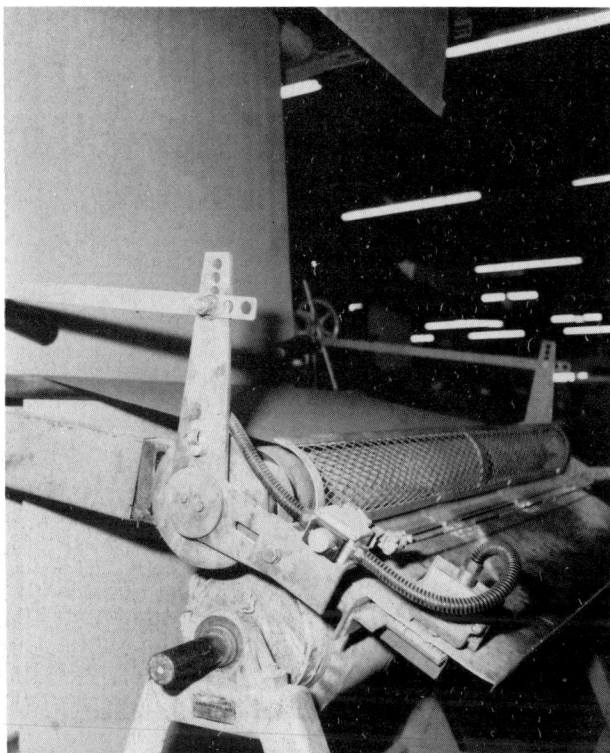

Fig 32 Self adjusting guard incorporating interlocked trip plate at A-frame batcher

(a) close fitting nip bars, even if notionally self adjusting, are not suitable. Few mechanisms have been devised to self adjust with sufficient accuracy (± 3 mm) to ensure safety, and a dangerous intake is thus readily created between the batch and the nip bar itself. Serious injuries have been caused in this way.

(b) a trip or pull wire is not effective on its own. Experience shows that a person being drawn into the intake is unlikely to have the presence of mind to operate the wire in time to avoid injury.

(c) altering the direction of rotation of the pressure roller and thus the batch makes the intake less accessible but is not an effective safety measure. Serious accidents have occurred at batches re-threaded in this way since approach to the intake is still possible.

Winding

154 Winding and rewinding of yarn into suitable packages such as cheeses, cones and and cops is a process ancillary to yarn bleaching and dyeing.

155 On modern winding machines relatively low noise levels can be achieved, but levels above 90dB(A) remain as a problem on traditional winding frames. In such cases wearing of personal hearing protection continues to be necessary until the noise can be adequately controlled by engineering means.

156 Where cotton yarn is processed, periodic dust surveys should be carried out to ensure that there is compliance with the occupational exposure standard for cotton dust, currently 0.5 mg/m³ total dust less fly, measured by static sample. See HSE Guidance Note EH25 *Cotton dust sampling*.[39]

DYEING PROCESSES

Hank dyeing

157 Hank or skein dyeing is a traditional technique method now only used for dyeing certain bulky yarns such as wool and acrylic for outerwear, knitwear and carpets. The process is tending to be superseded by package dyeing which has a higher productivity and takes up less floor space.

158 There are two variations of the process. In one, the hanks are suspended from a set of horizontal rods carried on a frame and are lowered into a rectangular vat of dye liquor, the liquor being circulated by a pump. In the other, the hanks are hung from perforated arms through which the dye liquor is pumped. The liquor runs down the skeins into the trough below and is recirculated.

159 The rods or perforated arms are periodically rotated to index the skeins, but there are no significant mechanical hazards apart from transmission drives which should be guarded in the usual way. The main danger is that of falling into the vat or trough of hot liquor so the rim of the vessel should be at least 920 mm (3 ft) above the adjacent platform or floor.

High temperature dyeing

160 High temperature dyeing is a general term for dyeing at elevated temperature and pressure (above 100°C and above one atmosphere). It includes beam, garment, jet and package dyeing unless such processes are carried on at

Fig 33 Combined door restraining and seal breaking device on multi-bolted door of high temperature dyeing machine

Fig 34 Pressure interlock on quick opening lid of high temperature dyeing machine

atmospheric pressure. Most high temperature machines can be used for preparation processes such as scouring and bleaching as well as for dyeing.

161 The main risks are if the access door or lid is opened under the following circumstances:

(a)   when the level of the liquor is above the rim of the door or lid, resulting in heavy spillage of liquor;

(b)   when the liquor is pressurised;

(c)   when the indicated temperature is less than 100°C but local hot spots are at or above boiling point (this can happen following interruption of the liquor flow as a result of fabric blockage or power failure).

In the case of lidded vessels, there is a further risk of falling into the liquor when the lid is open.

162   These risks continue to cause very serious and sometimes fatal scalding accidents. Control measures to be provided include:

(a)   a high standard of training, instruction and supervision of operatives;

(b)   safe systems of work covering loading, unloading and clearance of blockages, these systems of work to include the wearing of face shields and suitable protective clothing;

(c)   on machines with multi-bolted doors, a door restraint or a seal breaking device to allow only partial opening of the door until any residual pressure has been released;

(d)   on machines with quick opening doors, effective pressure and temperature interlocks, plus interlocks with the power and heating supplies;

(e)   on lidded vessels, the rim of the vessel to be at least 920 mm (3 ft) above the floor or working platform to reduce the risk of falling into the liquor.

163   Further information on the control measures and precautions needed is given in HSE Guidance Note PM4 *High temperature dyeing machines*.[25]

Fig 35 Electrical interlock on quick opening lid of high temperature dyeing machine

164 Proper maintenance of the complete machine is important, with attention being given to the integrity of the pressure vessel, the heating coil and the safety devices. Periodic thorough examination and testing should be carried out by a competent person with whom users should liaise to determine, for each individual machine, the necessary scope and frequency of the examination and test routines. Users should also make regular functional checks of the safety devices and keep a record of these checks on file.

Beam dyeing

165 Beam dyeing is a process suitable for lightweight woven and knitted fabrics, and is also used for dyeing warp beams.

166 The material, wound onto a perforated beam shell, is loaded into a closed vessel and subjected alternately to an inward and outward flow of liquor.

Some machines are loaded vertically by overhead crane but horizontal units are now more usual with the beam of fabric or yarn carried on a trolley.

167 Depending on machine design, beam dyeing can either be a high temperature process or be carried on at atmospheric pressure. If the former case applies, see *High temperature dyeing* paragraphs 160 to164.

Garment dyeing

168 Dyeing of garments requires purpose built machinery, the various types in use including:

(a)   side paddle machines;
(b)   overhead paddle machines;
(c)   side loading rotary drum machines;
(d)   combined rotary dyers and tumblers;
(e)   high temperature machines.

In all cases, these machines are also suitable for scouring and bleaching which is carried out as appropriate prior to the dyeing process.

169 At side and overhead paddle machines the goods, sometimes held in net bags, are circulated through a trough of liquor by a slowspeed paddle wheel. Interlocking is not normally provided for the access doors or covers.

170 At side loading rotary drum machines, the doors on the outer casing incorporate electrical interlocks so that the power supply is isolated except when these access doors are fully closed and locked. The doors on the inner cage are secured by mechanical catches.

171 With combined rotary dyers and tumblers, there is usually a high speed centrifugal period at the end of the wet processing cycle when even balancing of the load is essential, controlled by a self-balancing system. There is also a requirement for a safety interlock on the outer access door so that:

(a)   the machine can only operate when the door is closed and locked; and

(b)   the door can only be unlocked when working parts of the machine are at rest.

A time delay or motion sensing device is necessary to satisfy this second condition.

172 For high temperature garment dyeing machines, see *High temperature dyeing* paragraphs 160 to164. In the particular case of high temperature toroid machines, if the liquor flow should cease on account of power or pump failure, the garments would sink to the bottom of the vessel, effectively stratifying the liquor into top and bottom sections. Thus, if a cooling cycle were initiated, the bottom section would cool (and the temperature probe indicate a temperature substantially less than 100°C) while the upper section could remain at or above boiling point. A safe system of work to include the wearing of a face shield and suitable protective clothing should be adopted to minimise danger when dealing with such circumstances. A second, upper temperature probe or interlock would also enhance safety.

Jet dyeing

173 Jet dyeing is a system for dyeing fabric in continuous rope form, usually at high temperature, up to 135°C. See *High temperature dyeing* paragraphs 160 to 164. The action on the fabric is comparable to that of a winch but is carried on within a pressure chamber with the fabric propelled by a jet of liquor through the internal guide tube. The process, orignally introduced to dye polyester fabric, has been adapted to meet the varying requirements of other materials. Machine types include fully flooded, jet reel and soft flow machines.

174 A risk especially prevalent with jet dyeing and which has led to serious and fatal scalding accidents is opening the access door when the liquor temperature is too high. Such incidents are often associated with the clearing of fabric blockages which can occur inside the machine.

175 Following a blockage and the subsequent rapid cooling cycle which would be initiated by the operative, pockets of very hot liquor at or above boiling point may remain near the blockage even though the temperature gauge is measuring a temperature substantially less than 100°C. In these circumstances, if the door were opened and the fabric near the blockage disturbed, the operative could be engulfed by an eruption of steam and scalding liquor. To overcome this risk, a safe system of work for clearing blockages should be drawn up and enforced.

176 Essential elements of the system of work are:

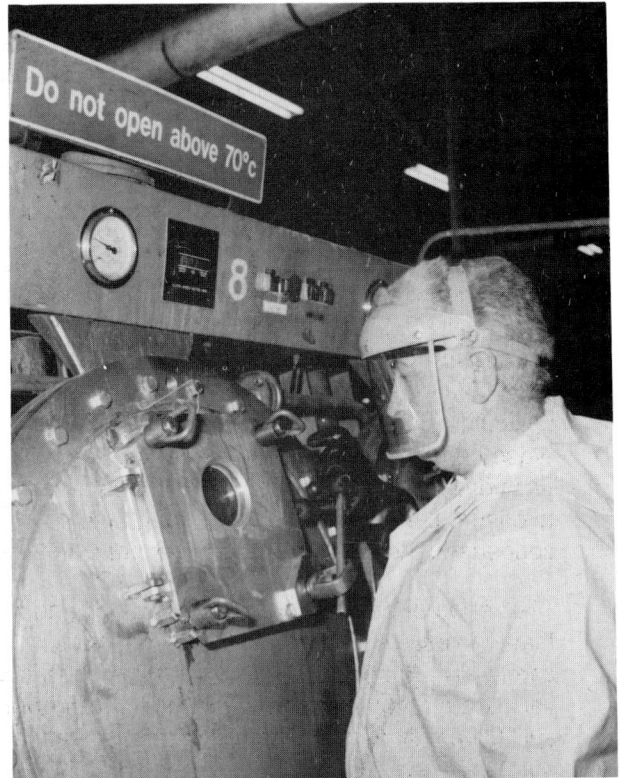

Fig 36 System of work for clearing blockages at jet dyeing machine

(a) cool to an indicated temperature not exceeding 85°C, or not exceeding any lower temperature specified by the machine manufacturer;

(b) put on protective equipment to comprise, as a minimum, a face shield plus impervious apron, gauntlets and footwear;

(c) re-check that the indicated temperature remains less than 85°C or less than any lower temperature specified by the machine manufacturer;

(d) standing to the side, open the access door;

(e) remain standing to the side and slowly withdraw the fabric from the blockage using a fabric hook if necessary.

Package dyeing

177 Most present day package dyeing machines for dyeing yarn and loose stock are pressurized, operating at temperatures of up to 135°C. See *High temperature dyeing* paragraphs 160 to164. The packages are mounted onto a carrier (or in the

case of loose stock, loaded into a cage) and lifted into the pressure vessel by overhead crane. The vessel lid is closed and locked, and the packages subjected alternately to an inward and outward flow of liquor. Modern developments in the process are leading to automatic loading and unloading of the packages.

178 During operation of the machine, an hydraulic lock and thus serious overpressure would be possible if the liquor bleed line became blocked for any reason. This has been known to happen particularly with loose stock dyeing, hence the necessity for a properly set and maintained pressure relief valve. In addition, at the start of each dyeing cycle, a visual check should be made to ensure that liquor is flowing through the condenser and bleed line into the expansion tank.

179 Another problem with loose stock dyeing is that the vessels tend to over-flow if operated carelessly. Flanged lids and remote steam shut off valves are recommended to reduce the risk from spilt liquor. Automatic control can also enhance safety.

Jigs

180 Dye jigs or jiggers, used for processing fabric in open width, comprise two power driven draw rollers positioned over a bath. The fabric is repeatedly transferred from one draw roller to the other, passing by means of guide rollers back and forth through the bath. Although classified as dyeing machines, jigs are also used for preparation treatments prior to the dyeing process itself. Most machines operate at atmospheric pressure although a few pressure jigs comprising a jig within a pressure vessel have been built.

181 The bath of liquor is heated by steam, a dual system generally being employed. Firstly, direct injection of live steam is used to reach operating temperature, then a closed steam coil maintains the desired process conditions. During the direct injection of live steam high noise levels are generated, up to 100 dB(A) being commonplace for short periods. These noise levels can be substantially reduced and thermal transfer efficiency improved by causing either liquor or air to be entrained with the flow of steam. See *Noise* paragraphs 68 to 71 for details of these straightforward noise control measures. Keeping the top covers in position also helps to control the noise problem and prevents excessive escape of steam into the general atmosphere of the dyehouse.

182 Jigs are loaded by transferring fabric from a batching roller onto one of the draw rollers. The batching roller is frequently carried on a square ended shaft and guarding is needed to avoid risk of entanglement on the exposed shaft ends. There is also a risk of entanglement where the fabric runs onto the draw rollers, during both loading and normal operation. Some machines incorporate a compensation device which monitors the fabric tension and switches off the machine automatically if excessive tension is detected for any reason.

Winches

183 Winches, also known as a becks or vessels, comprise a large diameter winch reel together with a smaller diameter jockey roller located above a bath. Several loops of fabric in rope form and arranged side by side are pulled round by the winch reel, the fabric thus passing repeatedly through the liquor in the bath. Desizing, scouring and bleaching can be carried out as well as dyeing.

184 The chemical hazards of the process and the risk of scalding are the main dangers. To minimise the risk of a person falling in, the top edge of the bath should be at least 920 mm (3ft) above the floor or working platform.

185 Winches are entered occasionally for maintenance purposes. Entry should be restricted until liquor and steam valves have been securely locked off and the temperature has reduced to a safe level.

Continuous dyeing

186 Continuous dyeing of fabric can be considered in two parts:

(a) application of the dye solution;

(b) fixation of the dye, possibly followed by washing off of surplus dye and auxiliaries.

187 Application of dye solution is normally by pad mangle although other methods may be used, for example by spraying or by the application of foam. Fixation may be in steam, hot air or by oxidation.

188 Plant items likely to be encountered in continuous dyeing are:

(a) washers;
(b) pad mangles;
(c) steamers;
(d) dryers;
(e) fabric take-off units.

189 The size of the plant may cause communication difficulties between operatives. This problem should be anticipated and systems of work adopted to minimise danger. These systems of work should include the use of audible or visual warnings when the plant is about to be started and a tannoy system may be justified in some cases. Suitably placed stop buttons should also be provided to protect operatives when away from the main control panel.

PRINTING PROCESSES

Flat screen printing

190 Flat screening printing is still carried on as a hand process especially for the printing of tea towels but in production volume terms has been largely superseded by fully automatic flat bed machines.

191 The cycle of such machines is as follows:

(a) the fabric, gummed to an endless backing blanket, is indexed forward and stops;

(b) the screens, one for each colour, drop down onto the fabric;

(c) the squeegee blades or rods move over the screens, pressing the printpaste through the mesh onto the fabric;

(d) the screens lift, the fabric is indexed to the next print station and the cycle continues.

192 As the printed fabric leaves the machine it is separated from the endless blanket and passes into a drying or curing oven. Meanwhile the blanket, possibly following washing and drying, returns to the feed-in end of the plant.

193 A significant risk arises from the motion of the hydraulic ram and the mechanical linkages beneath the table which index the blanket forward. A safe system of work should be followed to ensure that the side guards are fully secured in

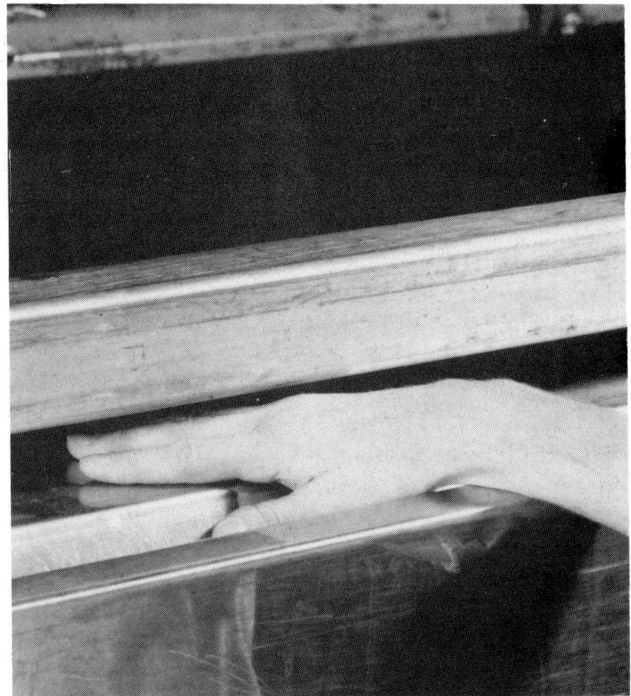

Fig 37 Safe gap below screen lifting bar at flat screen printing machine

position to prevent access to the underside unless the machine is stopped and isolated. Any roller intakes also require guarding.

194 There may be a finger trapping risk between the screen lifting bars and the machine table. This risk can be avoided by ensuring an adequate clear gap beneath the bars when they are in the fully down position.

195 Most of the printpastes are waterbased but where white spirit is used as the solvent a high standard of ventilation is needed to control the fumes given off. At the dryer, explosion relief is required at all direct gas fired units or where the use of white spirit gives rise to a solvent vapour explosion risk. See *Bakers curers and dryers* paragraphs 231 to 234.

Rotary screen printing

196 With rotary screen printing, colour is forced through the rotary screens and onto the fabric by means of a squeegee blade acting on the inside of each screen. In many respects the process has largely superseded engraved roller printing. It is particularly suited to wide fabrics, up to 3 m (10 ft) being quite normal, and short production runs are economical.

Fig 38 Access platform at rotary screen printing machine

Fig 39 Exhaust ventilation at laser engraver

197 Depending on the design of the individual machine, mechanical hazards which may require attention include:

(a) the intakes between the gearwheels on the machine bed and the driven gears at the screens;

(b) the horizontal driveshafts which run parallel to the screens.

198 So that the machine can be stopped immediately in the event of an emergency, trip wires are recommended in front of and parallel to each screen. On wide machines where it is not possible for a person to reach across to the mid point of the screens, a suitable walkway should be provided to give safe access. Obtaining access by means of the horizontal bars which carry the screens is not acceptable.

199 The screens are made from very fine nickel mesh and can cause severe cuts to the hands if the mesh is damaged. Suitable gloves should be worn if it is necessary to handle damaged screens. Cuts can also be caused by squeegees which incorporate sharp metal blades especially on wide machines where the long squeegees are difficult to handle. In many cases, gloves are not a practical solution to this problem. Some makers can supply squeegees edged with plastic or rubber rather than metal. These are intrinsically less dangerous and are recommended where technically feasible.

200 Screen cleaning involves the use of high pressure water jetting and care should be taken to avoid injecting water into the skin. Where chemical cleaning of the screens is also necessary, the chemical and fire hazards which may be introduced should be assessed.

Screen making

201 Both flat and rotary screens are sometimes made and repaired on site at the printworks. Any solvent vapour hazards arising from the adhesives and varnishes used should be adequately assessed with control measures applied as necessary. With rotary screens, if epoxy adhesive is used to secure the screens to their end rings there is a skin contact risk until the adhesive is fully cured.

202 Ultra-violet light is used to cure the light sensitive lacquer on flat screens to produce the desired print pattern. The light source should be suitably shielded or enclosed, using dark coloured curtains for example, to avoid unnecessary exposure to ultra-violet light. The controls for the light source should be positioned outside the enclosure.

203 Laser engraving of rotary screens is a technique now coming into more common use. The main problems are electrical safety and risk of injury to the eyes. Shields and covers should be

31

effectively interlocked to prevent access until the laser is de-energised. At the engraving head, exhaust ventilation should be applied to remove the fumes given off.

Transfer printing

204 Transfer or sublistatic printing is a process where fabric is heated and held in close contact with a printed paper, printed with disperse dyestuffs. The paper heats up causing transfer, by sublimation, of the dyestuffs from the paper to the fabric. The process is completely dry and the fabric, once printed, requires no subsequent steaming or washing. The process is particularly suitable for synthetic fabrics or garment panels.

205 For woven fabrics, the machine comprises a rotating cylinder up to 2 m (6ft 6in) in diameter together with an endless blanket which grips the printed paper, the fabric and a backing paper and takes these materials round the cylinder. A surface temperature of up to 220°C is required and the cylinder is heated internally either electrically or by circulating thermal fluid.

206 The main hazards are roller intakes, especially at the point where the blanket runs onto the cylinder, and a close fitting fixed guard is required at this point. An infra-red heater may be used to preheat the fabric to speed up the process. If the fabric stops, the heater should shut down automatically to prevent the risk of fire.

207 For knitted fabrics, the cylinder is quite small with heating provided by infra-red panels around part of its circumference. For garment panels, presses with steam heated heads are used, similar in principle to conventional garment presses. The main risk is burns from contact with the heads.

Roller printing

208 Engraved roller printing machines comprise:

(a) a large, central cylinder around which passes the fabric to be printed (the fabric is backed by a blanket and usually by a back grey fabric to give resilience during printing);

(b) a number of engraved printing rollers (each with its colour box, colour furnishing roller and doctor blades) positioned at the periphery of the cylinder, each contributing one colour to the fabric.

Fig 40 Remote pitching device at roller printing machine, side guard removed for clarity

209 The printing rollers are positively driven by gearing from the central cylinder. As the cylinder rotates (bringing with it the fabric, the blanket and the back grey) the rollers rotate at the same surface speed and transfer colour to the fabric.

210 The colour furnishing roller and the two doctor blades associated with each printing roller work as follows:

(a) the colour furnishing roller, rotating partly immersed in its colour box, picks up printpaste and transfers it to the printing roller;

(b) the first doctor blade clears excess colour from the printing roller, leaving colour only in the engraved pattern;

(c) the printing roller, pressing onto the fabric, transfers colour from the engraved pattern to the fabric;

(d) the second doctor blade (the lint doctor) clears the outrunning side of the printing roller of any lint or loose fibre picked up from the fabric.

211 After printing, the fabric is dried on cans and is led away for steaming and ageing. The back

Fig 41 Storage rack for doctor blades

Fig 42 Exhaust ventilation at plating bath

grey is taken off and later washed prior to re-use while the endless blanket normally passes through a blanket washer and dryer before returning to the input side of the machine.

212 Roller printing machines, as well as printing directly onto white cloth, can be used for both discharge and resist printing, these latter processes incorporating a padding operation. Padding takes place before printing of the discharge, after printing of the resist. Additional pad mangle risks are introduced but the hazards of the printing process itself remain unchanged. These hazards are principally:

(a) *gearwheel intakes at the crown and box wheels*. Where a remote pitching device is provided to give adjustment to the register of the rollers, there is no difficulty in achieving full enclosure of the crown and box wheels. Where a remote pitching device is not provided and adjustment is by means of a tommy bar acting on a screw, the crown and box wheels should be enclosed to the greatest practicable extent.

(b) *intakes between the printing rollers and printing cylinder*. These should be protected by fixed guards or nip bars unless adjacent doctor blades effectively prevent access.

(c) *doctor blades*. These are extremely sharp and can cause deep lacerations to the hands. A blade guard or cover should always be in position except when the blade is actually mounted on the printing machine or when it is being re-ground in the workshop. Purpose designed racks should be provided for storing the blades when not in use.

213 A specified procedure should be followed where manual cleaning is carried out. The machine should be stopped, guards removed as necessary and accessible parts cleaned. The machine should then be inched to expose other parts of the cylinder and rollers for cleaning, and so on. Alternatively the machine may be driven at a slow crawl not exceeding 5 metres/minute. No attempt should be made to clean rotating parts using a rag or similar material as this can get caught in the nip and draw in the operative's hand.

214 Manufacture of the printing rollers now tends to be carried out by specialist engravers rather than at the printworks. Aspects of the manufacturing process needing attention include copper plating, chromium plating, acid etching and use of solvents. A high standard of exhaust ventilation is required where harmful fumes are emitted.

FINISHING PROCESSES

Coating and laminating

215 Coating of fabric with various surface finishes is becoming increasingly common. Some of the chemicals used are water based but there may still be health hazards from fumes given off, and an associated fire or explosion risk at the subsequent drying stage, if any, in an oven or curer.

216 Lamination processes include adhesive lamination of fabric, and flame bonding of expanded foam plastic onto fabric. Effective exhaust ventilation should be provided at both processes, to control respectively the adhesive fumes and the isocyanate fumes given off. Where there is also a high fire risk, an integral fire fighting system is necessary.

217 Any intakes at rollers or mangles require guarding in the usual way. At flame bonding, the burner bars may restrict access to the intakes but the plant can be run cold with the burners out of position so additional guarding should be provided to prevent danger. Hinged or rise and fall interlocked screens are suitable.

218 Some gas burners used at flame bonding are very noisy, emitting a high pitched screech. Quiet burners are readily available and should be fitted.

Flocking

219 The flocking process, producing for example thermal linings for curtains, may incorporate a high voltage electrostatic unit and generally operates as follows:

(a) adhesive is applied by roller to the fabric;

(b) the electrostatic unit charges the flock particles, causing then to be raised on end and attracted to the fabric where they are bonded by the adhesive;

(c) the fabric passes on through an oven to cure the adhesive, thus completing the process.

220 The main problems requiring attention are:

(a) roller nips;

(b) fumes and possible fire risk from the adhesive;

(c) fire risk from the flock particles;

(d) high voltage electricity.

221 To control the electrical risks at the electrostatic unit, where fitted, the following precautions are needed.

(a) a full height enclosure should be provided to prevent access to the flocking area;

(b) the doors to the enclosure should be effectively interlocked so that opening of any door disconnects the power supply and automatically earths the high voltage conductors. Relying on manual earthing is not sufficient.

Stenters

222 Final treatment of fabric is usually carried out on the stenter frame which in combination with an oven controls the finished density and width of the fabric and has a significant effect on its appearance, handle and other properties.

223 The fabric, gripped at its edges by pins or clips carried on two endless chains, is conveyed by these chains through the oven compartment where circulating hot air completes the finishing treatment. On leaving the oven, the fabric is disengaged from the chains by a take-off roller and passes onwards through draw rollers to either an A-frame batching unit or plaiter.

224 Guards should be provided where the chains run onto the main drive sprockets and are also needed for any gears located beneath the chains. On older plant where drive shafts may run the length of the stenter, these drive shafts also require guarding.

225 Fabric entry to the stenter is controlled by a series of guide, feed and scroll rollers, the precise arrangement of which varies with different makes of stenter. Serious accidents have occurred at the entry zone with hands and arms being taken in by powered and non-powered rollers, or between fabric and roller. If entanglement does occur, many strong fabrics fail to tear, thus increasing the risk of injury. Fixed or interlocked guards should be provided to prevent access to the rollers.

226 At some stenters, automatic weft straightening is carried out to correct distorted

Fig 43 Sheet metal guard where stenter chain runs onto drive sprocket

Fig 44 Rise and fall guard for fabric entry zone at stenter

fabric. The weft straightening unit, usually positioned upstream of the fabric entry zone, comprises an assembly of skewed and barrel rollers. These rollers are not positively driven, being rotated only by the motion of the fabric. Nevertheless, serious entanglement accidents have occurred between such rollers and the fabric, and guarding is required. Sliding, interlocked guards at front and rear have proved successful.

227 Where knitted fabric is processed, it is common practice at the delivery end of the stenter to slit the fabric edges prior to batching. Guarding for the circular slitting knives should be effective when the knives are in both their working and raised positions. Automatic cross-cutting of the fabric is also a feature of some machines. A degree of protection can be achieved by incorporating a pressure switch on the stenter platform so that cross cutting can only take place when the operative is on the platform away from the danger zone.

228 Heating of the stenter oven is usually by steam, oil (indirectly through a heat exchanger) or gas, the latter now being the more common

method. Operating temperatures are typically in the range 60°C to 220°C, with the higher temperatures tending to give a greater fire risk. Regular cleaning of the oven and associated ductwork should be carried out to reduce the risk from combustible residues.

229 With **direct** gas fired stenter ovens where the products of combustion enter the oven space there is a significant gas explosion risk. In all such cases, adequate explosion relief panelling should be fitted where technically feasible irrespective of the sophistication of the gas controls. See *Fire and explosion* paragraphs 42 to 45. If the plant stops or the ventilation drops below a predetermined value the gas supply should be turned off automatically. A qualified industrial gas engineer should be consulted on the control features needed for a particular application.

230 At the oven entry and exit slots, careful design and control of the air recirculation system is required to reduce the escape of unpleasant, harmful fumes into the work area. Air curtain systems are available and should be installed where necssary.

Bakers curers and dryers

231 Various bakers and curers, less elaborate than stenter ovens, are used for drying and curing at temperatures up to about 180°C. If there is risk of a solvent vapour explosion or if the unit is **direct** gas fired, adequate explosion relief panelling should be fitted where technically feasible. See *Fire and explosion* paragraphs 42 to 45. Appropriate safeguards should also be incorporated with the gas controls and a qualified industrial gas engineer should be consulted on the features needed for a particular application.

232 Drying or pre-drying at lesser temperatures may be required at other stages of the finishing process. Several drying techniques are available, see *Drying cylinders*, *Vacuum slots*, *Hydro-extractors*, *Dielectric heating* paragraphs 235 to 249.

233 Where direct contact heating such as drying cylinders is unacceptable or where hot flue heating is too slow, infra-red dryers may be used. These incorporate ceramic radiants or heated metal plates. In the case of direct gas fired infra-red dryers which are enclosed to conserve heat, explosion panels are needed or the access doors should be fitted with explosion catches and loose restraining chains.

234 Brattice and tubular dryers and perforated suction drum dryers are used for processing knitwear and hosiery. Drum dryers operate at relatively slow speeds at temperatures in the region of 130 to 140°C. There is risk from internal moving parts, in particular the intakes between the drums, and the machines should only be operated when all side doors are closed and fastened.

Drying cylinders

235 Drying cylinders, also known as cans or tins, are extensively used for the drying of fabric by surface contact. The cylinders, arranged in either vertical stacks or horizontal rows, are usually heated by steam although oil is sometimes used. Steam heated cylinders are steam receivers subject to periodic thorough examination by a competent person.

236 Intakes can be created between adjacent inrunning cylinders and the preferred solution to

Fig 45 Side guards at drying cylinders

this problem is to position the cylinders to give a daylight gap of at least 100 mm (4 in) so that danger is minimised. If this is not feasible, which may be the case with some existing plant, side guards and nip bars are necessary. It may be thought that the heat keeps operatives away and in practice this is often the case, but cylinders are sometimes run cold for partial or pre-drying, and there will be need for approach when threading up if the leader tape has been lost.

237 The Jetcyl dryer is effectively a modified set of drying cylinders where air is directed at the fabric to supplement the contact heating effect. A full enclosure is normally provided to reduce energy losses but the potential intake hazard between the cylinders remains.

Vacuum slots

238 Vacuum slots are pre-drying, suction dewatering units now sometimes used as a substitute for pre-drying on cylinders. The main problem is noise from the high power vacuum pump which should be enclosed, positioned or silenced so that the noise level in the work area does not exceed 90dB(A), and preferably does not exceed 85dB(A).

Hydro-extractors

239 Hydro-extractors comprise a perforated cage which is rotated at high speed within a casing and rely on centrifugal force for the removal of water from the textile goods in the cage. The main dangers are out of balance forces due to uneven loading which could result in disintegration of the machine, and risk of entanglement with the rotating cage and contents.

240 To overcome the entanglement risk, the access cover or lid for the machine should be provided with an interlock such that:

(a) the cage cannot be set in motion until the cover is closed; and

(b) the cover cannot be opened until the cage has come to rest.

241 To satisy the second condition the interlock should incorporate a suitable time delay or a rotation sensing device. A possible design for the latter is illustrated at figure 74 of BS 5304: 1988 *Safety of machinery*. The interlock will tend to wear with use and regular testing is necessary.

242 The foundations of the machine should be properly anchored to the floor and balanced loading of the cage is essential. All operatives should be fully aware of this requirement although modern machines incorporate movement sensors to prevent starting of uneven loads.

243 To prevent overstressing of the cage, its safe working load should be established in terms of the material processed. Periodic thorough examination and non-destructive testing of the cage is also advised in line with the recommendations contained in BS 767: 1983 *Specifications for centrifuges of the basket and bowl type for use in industrial and commercial applications*.[43] Particular attention should be paid to the basket condition and to the lid fastenings and interlock. Records of the examinations and tests should be held on file.

244 The very small, domestic type of hydro-extractor (spin dryer) may be found in use, perhaps in the laboratory or quality control department. In cases where the momentum of the basket and contents is not high, and where operative training and supervision is good, an interlock on the top cover which merely isolates the power supply when the cover is opened may be sufficient to ensure safety. However if safety is in any doubt, a time delay or a rotation sensing device should be provided as described at paragraphs 240 and 241.

Dielectric heating

245 Many non-metallic materials (dielectrics) heat up uniformly when subjected to high frequency electromagnetic fields. The reason is that as the direction of the field reverses rapidly, so does the polarization of individual molecules thus causing internal friction and heat.

246 Water is very receptive to microwave and radio frequency (RF) energy whereas textile fibre is less so, making dielectric heating efficient for water removal from textile materials. The technique is attractive in that it is possible to dry uniformly and accurately to the correct moisture content although to save on energy costs, pre-drying by more traditional methods is usually carried out. RF heating tends to be employed in preference to microwave heating and is increasingly used for drying yarn packages and hanks.

247 The main risks from RF heating are:

(a) electric shock;

(b) RF burns from the generator and applicator system;

(c) exposure to stray RF energy.

248 The RF unit should be enclosed and provided with locked panels or interlocks to prevent electrical danger and danger from RF live parts. A visual warning is recommended to show when the plant is energised. The possible problem of stray RF energy should be assessed with monitoring of power densities and field strengths carried out by a person competent in RF techniques, especially after any work which could have affected the efficiency of the shielding.

249 Further advice is contained in HSE Guidance Note PM51 *Safety in the use of radio frequency dielectric heating equipment*.[30]

Calenders

250 Calenders are used for modifying the performance and properties of fabric and comprise two or more pressure rollers or bowls usually arranged in a vertical stack. The bowls are of steel or cast iron, or have a resilient covering. The steel or cast iron bowls may be run hot, up to 220°C.

251 The bowls achieve the desired fabric characteristics using a combination of pressure, temperature and in some cases friction caused by differential surface speeds. Specially engraved bowls are needed for processes such as Schreinering and embossing which alter the lustre of the fabric.

252 At all calenders, guarding is essential for the intakes between adjacent inrunning bowls and in many ways the dangers are comparable to those at mangles. See *Mangles* paragraphs 109 to 114. However, at calenders, unlike at mangles, the lead-in of the fabric to the bowls is usually not at right angles to the vertical plane of the bowls so there should be little difficulty in providing and maintaining strong, close fitting nip bars or plates. This is the recommended guarding solution so long as nip bars of other than circular cross section are used.

Compressive shrinking

253 Compressive shrinking ranges shrink fabric by a small amount to give dimensional stability and to eliminate the stretch which may have been imparted earlier in the finishing treatment. Tradenames for varients of the process include Rigmel, Evaset and Sanforize. In the knitting sector, compressive shrinking is usually referred to as compacting.

254 There have been a number of serious accidents at shrinking machines, major risks being the intakes created by the compressive shrinkage belt or rollers, and by the endless drying felt at the Palmer drying section. Interlocked guards, or close fitting fixed guards or nip bars should be provided as appropriate. The compressive shrinkage belt has to be dressed periodically using a portable grinding attachment. When this is being done, purpose designed temporary guarding is needed for the intake created between the belt and the grinding attachment.

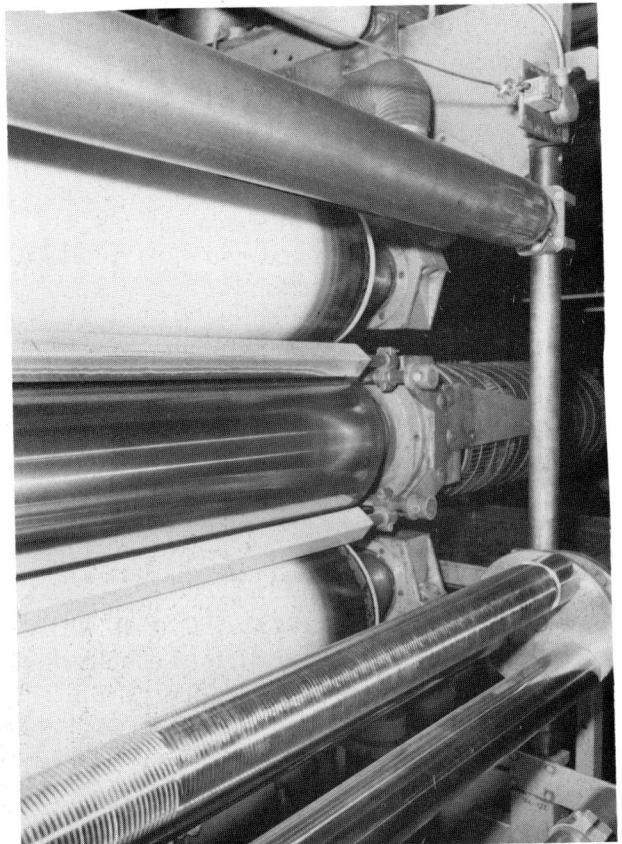

Fig 46 Square section nip bars at calender

255 The steam heated cylinders on the shrinking range are steam receivers subject to periodic thorough examination by a competent person.

Raising

256 Raising machines are used to brush and raise the surface of fabric to give it a warmer, fuller feel. They can be operated singly, or in tandem if both sides of the fabric are to be treated.

257 The raising effect is achieved by repeatedly passing the fabric, as an endless piece, over a large diameter, cylindrical abrading surface which rotates at high speed. The abrading surface is known as the barrel or drum and comprises an assembly of small diameter worker rollers, clothed in card wire. The barrel imparts some motion to the fabric but more positive transport and tensioning is provided by feed and draw rollers located at the input and output sides of the machine.

258 A number of serious accidents have occurred at the feed and draw rollers and the danger is

38

Fig 47 Interlocked screen at rear of raising machine, exhaust trunking also shown

particularly severe when the rollers are arranged as inrunning pairs. In this case, fixed or interlocked guarding is essential to prevent access. On most machines, the rollers can be moved apart a small distance to permit easy threading up with the working parts at rest. Any guard override key which may have been supplied with the machine should be removed.

259　On some machines the feed and draw rollers are single rollers relying on the friction effect of the wrap to transport the fabric. Depending on the configuration of the individual machine, an intake can exist between the rollers and the surface of the barrel. Since the machines are reversible, such an intake can be created at the side which is normally outrunning. Improved guarding may not be feasible on older machines but on modern plant, current

practice is to provide rise and fall interlocked screens at both front and rear to prevent access to the barrel and to the feed and draw rollers. These screens should extend close to floor level so there is no possibility of a person crawling underneath. Braked motors are also provided so that the machine stops quickly if any guard is opened.

260　The card wire of the worker rollers needs to be dressed periodically. The safest way of doing this is in the workshop with the rollers removed from the machine. If done in situ, there is danger from adjacent worker rollers and thus a need for purpose designed, temporary guarding. Preferably only the roller being dressed should be in motion which is possible if the drives to the barrel and rollers are independent. In all cases the work should only be carried out by authorised, trained personnel.

261　Effective exhaust ventilation is needed at raising machines to remove the dust and flock which is generated. This is particularly important where cotton fabric is processed.

Suedeing

262　Suedeing machines abrade the surface of fabric as it passes over a set of power driven abrasive covered worker rollers. Pressure of the fabric against the rollers can be adjusted by screwing in or out by hand a set of pressure bars.

263　The accident rate at suedeing machines is relatively low and guarding is not normally provided for the worker rollers if the wrap of the fabric on these rollers and the pressure bars is insufficient to cause a serious nip hazard.

264　Effective exhaust ventilation is needed at suedeing machines to remove the dust and flock which is generated. This is particularly important where cotton fabric is processed.

INSPECTION AND MAKING-UP PROCESSES

Inspection

265　Fabric is usually inspected as it is drawn over an illuminated table and is then either wound into a roll or plaited down.

Fig 48 Stop button at re-rolling machine

Fig 49 Take-up rollers at creasing and lapping machine

266   With fabric wound into a roll, the drive can either be:

(a)   a pair of take-up rollers (this method is often used at machines for inspecting knitted fabric);

(b)   a positive drive from the central shaft of the roll;

(c)   the turning effect, due to friction, of the roll resting on two rollers driven in the same direction.

267   Where a pair of take-up rollers is used, guarding is needed to prevent access to the roller intake. Where the other winding methods are used, a hazard exists where the fabric runs onto the roll. This is a problem particularly at the start of a new roll where the end of the fabric has to be tucked in to take up the drive. Serious entanglement accidents have occurred at this point and to mimimise danger, the following precautions should be taken:

(a)   the controls for the machine should be suitably positioned and shrouded to reduce the chance of inadvertent starting;

(b)   the fabric should only be tucked in when the machine is at rest;

(c)   a suitably positioned trip wire or stop button should be provided so that a person working near the roll can stop the machine in the event of an emergency.

Rolling

268   Rolling and re-rolling machines are used to wind or rewind fabric batches. Often the machines are made in house to the user's own design.

269   Serious entanglement accidents can occur as the operative assists initial take-up of the fabric onto the central shaft or box. To minimise danger, precautions similar to those described at *Inspection* paragraphs 265 to 267 should be taken since the problem is the same.

Fig 50 Side guard and interlocked front guard at plaiting machine

Creasing and lapping

270 Creasing and lapping machines fold fabric down the centre and wind it into an oval shaped package.

271 On traditional machines, the drive for the fabric is a pair of take-up rollers which are loaded either pneumatically or by spring pressure. Although these rollers are located within the body of the machine giving some positional safety, additional guarding may be needed to restrict access.

272 Threading up of the fabric with the machine at rest can be accomplished easily by releasing the pressure on the take-up rollers to give a small clearance.

Plaiting

273 Plaiting machines are used to lay fabric into folds or plaits of regular length for convenience of subsequent packaging.

274 As the reciprocating plaiting arm moves, trapping points are created:

(a) between the arm and the side framework of the machine;

(b) between the plaiting knife carried by the arm and the fixed card bar.

275 Substantial guards should be provided at the sides of the machine to prevent access to these trapping points and the guards should also protect any exposed shafts or pulleys. Small apertures may be necessary in the guards so that the machine can be adjusted and are acceptable so long as there is no possibility of contact with moving parts. The guards should be suitably interlocked if they require frequent removal.

276 An interlocked front guard is needed for the card bar trap. On older machines the interlock is by means of a mechanical linkage to the striking gear for the transmission drive. On more modern machines, electrical interlocking is provided.

Heat sealing

277 Heat sealing machines are used for wrapping and sealing rolls of fabric in plastic film to prevent damage during subsequent storage and transportation.

278 The heated wire or rod which fuses and seals the plastic film may cause harmful fumes to be emitted. If so, the machine should either be located in a well ventilated area or local exhaust ventilation should be provided to keep the fumes out of the working environment.

279 On automatic and semi-automatic machines, mechanical hazards arise from the various moving parts and effective guarding systems should be provided as necessary to minimise danger. An enclosure incorporating interlocked access panels may be required.

ALPHABETICAL INDEX

280 Alphabetical index for each machine and process described in Part 2. The numbers shown are the corresponding paragraph numbers.

PART 3 : FURTHER INFORMATION

RELEVANT LEGISLATION

On sale at HMSO Bookshops

Acts

1   *Factories Act 1961* ISBN 0 10 850027 6

2   *Health and Safety at Work etc Act 1974*
ISBN 0 10 543774 3

Regulations and Orders

3   SR and O 1938 No 106 *Kiers Regulations 1938*

4   SR and O 1938 No 641 *Operations at Unfenced Machinery Regulations 1938*
ISBN 0 11 100230 3

5   SI 1970 No 535 *Abrasive Wheels Regulations 1970* ISBN 0 11 000535 X

6   SI 1974 No 1681 *Protection of Eyes Regulations 1974* ISBN 0 11 041681 3

7   SI 1980 No 1471 *Safety Signs Regulations 1980* ISBN 0 11 007471 8

8   SI 1985 No 1333 *Ionising Radiations Regulations 1985* ISBN 0 11 057333 1

9   SI 1988 No 1657 *Control of Substances Hazardous to Health Regulations 1988*
ISBN 0 11 087657 1

10   SI 1989 No 635 *Electricity at Work Regulations 1989* ISBN 0 11 096635 X

11   SI 1989 No 1790 *Noise at Work Regulations 1989* ISBN 0 11 097790 4

12   SI 1989 No 2169 *Pressure Systems and Transportable Gas Container Regulations 1989*
ISBN 0 11 098169 3

HSE PUBLICATIONS

On sale at HMSO Bookshops

Legal guidance

13   *Factories Act 1961 : a short guide* 2nd ed 1977 ISBN 0 11 881111 8

14   HS(R)6 2nd ed 1983 *Guide to the Health and Safety at Work etc Act 1974* ISBN 0 11 883710 9

15   HS(R)7 *Guide to the Safety Signs Regulations 1980* ISBN 0 11 883415 0

16   Approved Code of Practice COP 16: *Protection of persons against ionising radiation arising from any work activity* ISBN 0 11 883838 5

17   Approved Codes of Practice L5: *Control of substances hazardous to health and Control of carcinogenic substances* ISBN 0 11 885593 X

18   HS(R) 25 *Memorandum of Guidance on the Electricity at Work Regulations* 1989
ISBN 0 11 883963 2

19   *Noise at Work: Noise Guides No 1 and No 2.* 1989 ISBN 0 11 885512 3

20   HS(R)30 *Guide to the Pressure Systems and Transportable Gas Containers Regulations 1989* ISBN 0 11 885516 6

21   Approved Code of Practice COP 37: *Safety of pressure systems* ISBN 0 11 885514 X

Plant and machinery safety

22   HS(G)16 *Evaporating and other ovens* 1981 ISBN 0 11 883433 9

23   HS(G)17 *Safety in the use of abrasive wheels* 1984 ISBN 0 11 883739 7

24   HS(G)37 *Introduction to local exhaust ventilation* 1987 ISBN 0 11 883954 3

25   Guidance Note PM4 *High temperature dyeing machines* 1980 ISBN 0 11 883049 X

26   Guidance Note PM5 *Automatically controlled steam and hot water boilers* rev 1989
ISBN 0 11 885425 9

27   Guidance Note PM22 *Training advice on the mounting of abrasive wheels* rev 1989
ISBN 0 11 883568 3

28   Guidance Note PM23 *Photo-electric safety systems* 1981 ISBN 0 11 883384 7

29   Guidance Note PM41 *Application of photo-electric safety systems to machinery* 1984
ISBN 0 11 883593 9

30   Guidance Note PM51 *Safety in the use of radio frequency dielectric heating equipment* 1986 ISBN 0 11 883615 3

31   *Deadly maintenance, plant and machinery : a study of fatal accidents at work* 1985 ISBN 0 11 883805 9

32   *Deadly maintenance : a study of fatal accidents at work* 1985 ISBN 0 11 883806 7

33   *Deadly maintenance, roofs : a study of fatal accidents at work* 1985 ISBN 0 11 883804 0

34   *Programmable electronic systems in safety related applications : an introductory guide* 1987 ISBN 0 11 883913 6

35   *Programmable electronic systems in safety related applications : general technical guidelines* 1987 ISBN 0 11 883906 3

Electrical safety

36   Guidance Note GS27 *Protection against electric shock* 1984 ISBN 0 11 883583 1

37   Guidance Note GS37 *Flexible leads, plugs, sockets etc* 1985 ISBN 0 11 883519 X

38   Guidance Note PM32 *Safe use of portable electrical apparatus* 1983 ISBN 0 11 883563 7

Chemical safety

39   Guidance Note EH25 *Cotton dust sampling* ISBN 0 11 883197 6

40   Guidance Note EH40 *Occupational exposure limits* (revised annually) ISBN 0 11 885580 8

41   Guidance Note EH 44 *Dust in the workplace : general principles of protection* ISBN 0 11 883598 X

42   *Guidelines for the safe storage and handling of non-dyestuff chemicals in textile finishing* 1985 ISBN 0 11 883833 4

BRITISH STANDARDS

Available from British Standards Institution, Linford Wood, Milton Keynes MK14 6LE

43   BS 767 : 1983 *Specifications for centrifuges of the basket and bowl type for use in industrial and commercial applications*

44   BS 1710 : 1984 *Specification for identification of pipelines and services*

45   BS 2092 : 1987 *Specification for industrial eye protectors for industrial and non-industrial use*

46   BS 2771 : 1986 *Electrical equipment of industrial machines, specification for general requirements*

47   BS 4275 : 1974 *Recommendations for the selection, use and maintenance of respiratory protective equipment*

48   BS 4293 : 1983 *Specification for residual current operated circuit breakers*

49   BS 4343 : 1968 *Specification for industrial plugs, socket outlets and couplers for AC and DC supplies*

50   BS 5304 : 1988 *Code of practice for safety of machinery*

51   BS 5490 : 1977 *Specification for classification of degrees of protection provided by enclosures*

OTHER GUIDANCE

Available from the addresses shown

52   IND(G)1(L) rev *Articles and substances used at work : the legal duties of designers, manufacturers, importers and suppliers, and erectors and installers* (free leaflet available from HSE Information Centres)

53   Health and Safety Executive 1980 *Safe handling of dyestuffs in colour stores* (available from HSE, West and North Yorkshire Area, 8 St. Paul's Street, Leeds LS1 2LE)

54   Institution of Electrical Engineers: *Regulations for Electrical Installations: IEE Wiring Regulations* 15th ed 1981 plus amendments (available from the Institution of Electrical Engineers, Savoy Place, London WC2R 0BL)

55   Knitting, Dyeing and Lace Industries Joint Health and Safety Committee : *Health and Safety Recommendations* (available from the Knitting, Dyeing and Lace Industries Joint Health and Safety Committee, 55 New Walk, Leicester LE1 7EB)